LIFELINES

CHEMISTRY FOR THE LIFE SCIENCES

■ Raul Sutton
School of Applied Sciences, University of Wolverhampton, UK

■ Bernard Rockett
Wolverhampton Grammar School, UK

■ Peter Swindells
School of Applied Sciences, University of Wolverhampton, UK

London and New York

First published 2000 by Taylor & Francis
11 New Fetter Lane, London EC4P 4EE

Simultaneously published in the USA and Canada
by Taylor & Francis Inc.,
29 West 35th Street, New York, NY 10001

Taylor & Francis is an imprint of the Taylor & Francis Group

Typeset in Perpetua 11/12 by Graphicraft Limited, Hong Kong
Printed and bound in Great Britain by TJ International Ltd, Padstow, Cornwall

British Library Cataloguing in Publication Data
A catalogue record for this book is available from the British Library

Library of Congress Cataloging in Publication Data
Sutton, Raul.
Chemistry for the life sciences / Raul Sutton, Bernard Rockett, Peter Swindells.
p. cm. – (Modules in life sciences)
Includes bibliographical references and index.
(alk. paper)
1. Chemistry. I. Rockett, Bernard W. II. Swindells, Peter. III. Title. IV. Series.
QD33.S93 2000 99–31650
540–dc21

ISBN 0–7484–0833–9

CONTENTS

SERIES EDITOR'S PREFACE

Teaching programmes in universities now are generally arranged in collections of discrete units. These go under various names such as units, modules, or courses. They usually stand alone as regards teaching and assessment but, as a set, comprise a programme of study. Usually around half of the units taken by undergraduates are compulsory and effectively define a 'core' curriculum for the final degree. The arrangement of teaching in this way has the advantage of flexibility. The range of options over and above the core curriculum allows the student to choose the best programme for her or his future.

The Lifelines series provides a selection of texts that can be used at the undergraduate level for subjects optional to the main programme of study. Each volume aims to cover the material at a depth suitable to a particular level or year of study, with an amount of material appropriate to around one-quarter of the undergraduate year. The concentration on life science subjects in the Lifelines series reflects the fact that it is here that individual topics proliferate.

Suggestions for new subjects and comments on the present volumes in the series are always welcomed and should be addressed to the series editor.

John Wrigglesworth
London, March 2000

INTRODUCTION AND HOW TO USE THIS TEXT

This book is intended as a self-study text for first-year undergraduate life scientists. These students are often expected to study some biochemistry modules during their course and will need access to selected fundamental chemical concepts in order to underpin this teaching. Thus, we have aimed to include only those aspects of chemistry that would be relevant to you, the life scientist. Each chapter begins with an introduction that explains the biological relevance of the contents. We have made a conscious effort to eliminate the overlap with conventional biochemistry textbooks. The relevance of the material to you is given importance by the choice of biologically relevant examples wherever possible. The text expects little in the way of prior knowledge of chemistry and introduces concepts at a basic level. However, the depth that the text seeks to achieve means that many concepts are introduced and expanded within a short space. This keeps each chapter brief whilst covering most of the necessary material that you will require during your undergraduate course. The text provides you with a concise introduction to chemistry.

Many undergraduate life scientists have not studied mathematics beyond GCSE and we have assumed little prior knowledge of mathematics when writing this text. Chemistry relies, in part, on physical principles, many of which are derived from fundamental considerations of equilibrium thermodynamics. The origin of equilibrium thermodynamics requires a reasonable grasp of algebra and calculus. We have removed such derivations from the text and presented only the equations that you will use when performing calculations. You may wish to satisfy your natural curiosity and so the derivations and origins of important equations are included in an appendix. We realise that you may not feel confident tackling problems of which you have had little prior experience. Each chapter helps you in areas involving calculations by illustrating each important type of calculation with step-by-step worked examples. You can then test your understanding of the steps involved by attempting the questions that are provided in the text. These in-text questions are supported by a set of fully worked answers so that you can check your approach to problem solving.

Some of the material covered in the chapters, such as the names and structures of functional groups, will require learning by rote and this is something that you will have to work towards. It would be impractical for you to remember all such names at the first attempt. Learning by rote can be made easier by breaking down the material into small parts. The regular review of such material will also aid your memory. It is similar with other ideas that are introduced in the text. Physical concepts can sometimes seem daunting to the life scientist but rereading material that you find difficult, slowly and carefully, will help you to gain a much clearer understanding of its meaning. In all areas of the book there are worked examples of answers to questions to help you to learn and in-text questions to help you to reassure yourself that you understand the part of the book that you are reading.

We use the term biomolecule freely throughout the text to denote a molecule that is an important constituent of living organisms. Many of the chemical reactions undertaken by living organisms take place in a controlled environment, often at or near to pH 7.0. Consequently, pH 7.0 and 25°C are assumed to be standard conditions for the purposes of this text.

The chemistry that you learn here will be used in many areas of your future study in subjects such as physiology, pharmacology, microbiology and biochemistry. A deeper understanding of biology can only come when we understand the structure, reactivity and physical processes of the molecules that make up the varied living world around us. This book will help you to appreciate this important area of the life sciences.

We wish to thank Dr John Edlin for permission to reproduce the Periodic Table used in Figure 1.1.

Raul Sutton
Bernard Rockett
Peter Swindells
May 1999

ELEMENTS, ATOMS AND ELECTRONS

1

■ 1.1 INTRODUCTION

In order to understand the nature and reactions of biological molecules which may have large or complex structures, it is helpful to examine first the simplest units of matter. This topic will show how a few elementary particles, protons, neutrons and electrons, can be used to build the atoms of elements and how the electrons are organised within atoms. This organisation determines the properties of the atom and how an element will be combined within an organism.

■ 1.2 MATTER AND ELEMENTS

Picture on one hand the darting flight of an iridescent dragonfly and on the other hand a pile of soil. The first is brilliant, dynamic, organised; the second is dull, featureless, inert. It does not seem possible that the two can be related in any way but we know that each is composed of the simple substances called **elements**. The 92 naturally occurring elements combine in a variety of ways to make up all the organisms and materials of the world we live in.

Each element is a single, simple substance; it cannot be split by chemical means

An element is a single substance that cannot be split by chemical means into anything simpler. Carbon is the building block of life; it is an element and cannot be split into anything simpler. Of the large number of natural elements only a few are of importance in the living world. Some assume major significance as macronutrients while others are required in small quantities as trace elements; these are shown in Tables 1.1 and 1.2.

In order to write biological substances in as clear and accurate a form as possible, the different elements or atoms can be represented in a brief, simple form called a **symbol**. Each element has a single capital letter as the symbol; thus carbon is represented as C and hydrogen as H. When more than one element begins with the same letter, then the symbol may have a second, small letter added to it to avoid confusion. In this way, calcium is shown as Ca, while chlorine is Cl. The conventional way of writing a symbol should always be used; chlorine cannot be written as CL or as cl. Symbols for biologically important elements are listed in Tables 1.1 and 1.2. An extended list is given in Figure 1.1.

A symbol is used as a shorthand form for an element

Table 1.1 Elements of major importance to plants and animals

Element name	*Symbol*	*Role in living organisms*	*Source used by man*
Carbon	C	Constituent of protein, carbohydrate, fat	Meat, fruit, vegetables
Hydrogen	H	Body fluid, essential for protein, carbohydrate, fat	Water
Oxygen	O	Essential for respiration, body fluid, protein, carbohydrate, fat	Air and water
Nitrogen	N	Constituent of proteins, nucleic acids, chlorophyll	Meat and fish
Phosphorus	P	Essential for ATP, phospholipids, nucleic acids	Meat and milk
Sulphur	S	Component of proteins, coenzyme A	Meat, fish, eggs
Chlorine	Cl	Ion balance across membranes, stomach acid	Table salt, salted foods
Sodium	Na	Ion balance across membranes	Table salt, salted foods
Potassium	K	Anion–cation balance across membranes, nerve impulses	Meat, green vegetables
Calcium	Ca	Component of bones, teeth, invertebrate shells, plant cell walls, essential for blood clotting	Hard water, milk

1.3 ATOMS

An atom consists of a nucleus made up of protons and neutrons, which is surrounded by electrons

Each element is made up of a large number of tiny but identical particles called **atoms**. Thus the element carbon is composed entirely of carbon atoms, while oxygen consists only of oxygen atoms. An atom is described as the smallest particle into which an element can be divided while still retaining the properties of the element. The very small size of a carbon atom can be appreciated when it is found that 12 g of carbon contain 6×10^{23} individual atoms; this value is the Avogadro number. It is important to us as it enables the masses of different elements to be compared. One atom is much too small to be weighed easily, so we take the mass of 6×10^{23} atoms of an element and call this the **atomic mass** of the element. Thus the atomic mass of carbon is 12 g. It is convenient to compare the atomic mass of an element to one-twelfth of the atomic mass of carbon and call the value obtained the **relative atomic mass** (A_r). The relative atomic mass of carbon is 12 (there are no units), the value for hydrogen is 1 and for oxygen is 16.

1.4 ATOMIC STRUCTURE

An electron in an atom is in rapid, random motion

Although an atom is the smallest chemically distinct particle of an element, it does consist of smaller units known as **subatomic particles**. There are three types of subatomic particles, **protons**, **neutrons** and **electrons**. The number and arrangement of these particles in the atom determine which element it is and how it will react in biological or chemical processes. Each proton carries the same positive electrical charge while the neutron is electrically neutral. These two particles have almost the same mass and are joined tightly together in the tiny central part of the atom, the **nucleus**. Each electron carries a negative electrical charge which exactly balances the positive charge on the proton. Electrons have a very small mass compared with protons and neutrons. They are distributed around the nucleus in discrete **energy levels** or **orbitals**. The three subatomic particles are compared in Table 1.3. The same three subatomic particles are present

Table 1.2 Trace elements of importance to animals and plants

Element name	*Symbol*	*Role in living organisms*	*Source used by man*
Boron	B	Healthy cell division in the growing points of plants	–
Fluorine	F	Constituent of teeth and bones	Hard water, milk
Iodine	I	Essential for thyroxine in thyroid	Drinking water, sea food, iodised table salt
Selenium	Se	Removal of active oxygen species by glutathione peroxidase	Fruit and vegetables
Manganese	Mn	Growth of bone	Present in a range of foods
Iron	Fe	Oxygen carrier in myoglobin and haemoglobin, cofactor in many reduction/oxidation (redox) reactions	Liver, red meat, spinach
Cobalt	Co	Essential for vitamin B_{12} promoting red cell development	Liver, red meat
Copper	Cu	Oxygen carrier in haemocyanin for certain invertebrates; component of cytochrome oxidase, an enzyme found in the respiratory chain of almost all eukaryotes	Present in a range of foods
Zinc	Zn	Essential in carbonic anhydrase for carbon dioxide transport in blood	Present in a range of foods
Molybdenum	Mo	Plant enzymes involved with nitrogen fixation and formation of amino acids	–
Silicon	Si	Cell walls in plants, exoskeletons of marine invertebrates	–

Table 1.3 Subatomic particles

Particle name	*Approximate relative mass*	*Relative electrical charge*
Proton	1.0	1+
Neutron	1.0	0
Electron	0.002	1–

in the atoms of every element but occur in different numbers and proportions in different elements.

The number of protons in the nucleus determines to which element the atom corresponds. For example, hydrogen always has one proton in each of its atoms, carbon has six protons and oxygen has eight protons. This number of protons in the nucleus of an atom of a given element is called the **atomic number** or **proton number**. An atom is

The atomic number of an element is the number of protons in each atom of the element

Table 1.4 Atomic structure and electron structure of isotopes of biologically important elements

Element name	*Symbol*	*Number of protons (atomic number)*	*Number of neutrons*	*Number of electrons*	*Full symbol*
Hydrogen	H	1	0	1	$^{1}_{1}H$
Deuterium	H	1	1	1	$^{2}_{1}H$
Boron	B	5	6	5	$^{11}_{5}B$
Carbon	C	6	6	6	$^{12}_{6}C$
Carbon	C	6	7	6	$^{13}_{6}C$
Carbon	C	6	8	6	$^{14}_{6}C$
Nitrogen	N	7	7	7	$^{14}_{7}N$
Oxygen	O	8	8	8	$^{16}_{8}O$
Sodium	Na	11	12	11	$^{23}_{11}Na$
Magnesium	Mg	12	12	12	$^{24}_{12}Mg$
Phosphorus	P	15	16	15	$^{31}_{15}P$
Sulphur	S	16	16	16	$^{32}_{16}S$
Chlorine	Cl	17	18	17	$^{35}_{17}Cl$
Chlorine	Cl	17	20	17	$^{37}_{17}Cl$

electrically neutral which means that the number of protons and electrons must be equal. Hydrogen has one proton and so has one electron, nitrogen has seven protons and therefore must have seven electrons. The atomic nucleus of the lighter elements often contains an equal number of neutrons and protons; carbon has six protons and six neutrons, oxygen has eight protons and eight neutrons. Heavier elements tend to have a greater number of neutrons than protons (see Figure 1.1). Hydrogen is unique in having no neutrons in the nucleus. The elements of biological importance are shown in Table 1.4 with the numbers of subatomic particles in the atom.

The mass number is the sum of the number of protons and neutrons in the element

The number of protons and neutrons in the atom and, by implication, the number of electrons, is represented in a shorthand form based on the symbol for the elements. The number of protons is shown below (subscript) and to the left of the symbol; this is the **atomic number**. The sum of the number of protons and neutrons is shown above (superscript) and to the left of the symbol; this is the **mass number**.

WORKED EXAMPLE 1.1

The element nitrogen, which is important in amino acids, has seven protons and seven neutrons in the nucleus of the atom.

(i) Show this in terms of the element symbol.

(ii) How many electrons are there in the atom?

ANSWER

(i) Write the symbol for nitrogen N

Add the number of protons (7) below and to the left of the symbol $_{7}\mathrm{N}$

Add the sum of the protons (7) and neutrons (7) above and to the left of the symbol $^{14}_{7}\mathrm{N}$

(ii) The nitrogen atom has seven protons. The number of protons equals the number of electrons in any atom, thus the nitrogen atom has seven electrons.

When the symbol is provided, then it can be used to find the number of neutrons and electrons.

WORKED EXAMPLE 1.2

Boron is a trace element which promotes healthy cell division in plants; it has the symbol $^{11}_{5}\mathrm{B}$. How many neutrons and how many electrons are present in the atom?

ANSWER

The number of neutrons is obtained by subtracting the atomic number (5) from the mass number (11): 11 – 5 = 6 neutrons.

The number of electrons is the same as the number of protons (the atomic number), 5; the boron atom has five electrons.

QUESTION 1.1

Sodium is an element involved in the ionic balance across membranes.

(i) How many electrons and neutrons are present in the atom?

(ii) Use Table 1.4 to write the symbol for the element to show the mass number and atomic number.

1.5 ISOTOPES

It has been shown that the atom of a specific element contains a fixed number of protons and electrons; hydrogen always has one proton and one electron, carbon invariably contains six protons and six electrons. However, the number of neutrons in the atom of an element can vary. Thus a small proportion of hydrogen atoms have one neutron rather than none and some carbon atoms have seven or eight neutrons, rather than the more common six neutrons. These different forms of the same element are called **isotopes**. They show the same chemical properties but have different mass numbers and thus different masses.

Isotopes of an element have different numbers of neutrons in atoms of the element

In the living organism, isotopes of an element are completely interchangeable with one another, although the heavier isotope reacts more slowly. The difference in speed (rate) of reaction between the two forms depends on the mass difference. Normal hydrogen ($^{1}\mathrm{H}$) has only half the mass of deuterium ($^{2}\mathrm{H}$) and so it reacts considerably faster. The symbol D often replaces $^{2}\mathrm{H}$. Biochemical pathways are investigated by replacing hydrogen with deuterium, perhaps by substituting deuterium oxide (heavy water D_2O) for ordinary water (H_2O). The fate of the deuterium in the pathway can be traced by routine analysis, hence the term 'tracer' is often used for isotopes such as deuterium.

Different isotopes of an element have different mass numbers

Isotopes of an element may be stable or unstable. Stable isotopes such as $^{12}_{6}C$ and $^{13}_{6}C$ do not undergo any change in the atomic nucleus with time while unstable isotopes such as $^{14}_{6}C$ (carbon-14) decompose spontaneously at a fixed rate. It is the atomic nucleus of the isotope that decays. Such behaviour is often associated with a high proportion of neutrons to protons in the nucleus. A neutron in the nucleus of an unstable isotope can break down to form a proton and an electron. The proton remains in the nucleus while the electron is emitted at high energy and can be detected with a suitable instrument. The high-energy electron is called a **β-particle** and this type of breakdown is known as **β-decay**.

Conversion of a neutron in $^{14}_{6}C$ to a proton and an electron leaves the nucleus with seven neutrons and seven protons: it becomes an isotope of nitrogen. The reaction can be represented:

$$^{14}_{6}C \rightarrow {}^{14}_{7}N + \beta\text{-particle}$$

It takes place quite slowly, with half of the original amount of carbon-14 breaking down over a period of 5760 years. The decay of unstable isotopes is known generally as **radioactive decay**. Radioactive decay does not always result in the emission of β-particles. Some isotopes emit α-particles (the nuclei of helium atoms, $^{4}_{2}He$), while others emit high-energy electromagnetic radiation (γ-rays) on decay.

QUESTION 1.2

Write an equation to represent the β-decay of the radioactive isotope $^{35}_{16}S$.

1.6 THE PERIODIC TABLE

It is convenient to arrange the elements in order of increasing atomic number. When this is done it is found that the properties of the eight light elements, lithium to neon, are repeated in the next eight elements, sodium to argon. A table arranged to align in vertical columns elements with similar properties is called a **periodic table** (Figure 1.1).

Only a few of the elements are of significance in the life sciences – these are highlighted in Figure 1.1. They are grouped mainly to the left (metals) and to the right (non-metals) with a third group in the middle (transition metals). The periodic table shows a number of trends, some of which are important in biological chemistry. **Electronegativity**, which is the power of an element to attract electrons to its atoms, varies in the periodic table. It increases on moving from left to right and from bottom to top. Thus the important biosphere elements nitrogen, oxygen, and to a lesser extent, sulphur and chlorine, are electronegative. This has consequences for the reactivity of biological molecules (Chapter 8) and for the formation of hydrogen bonds which leads to the shapes of protein molecules.

Electronegative atoms strongly attract electrons

The size of atoms of the elements decreases on passing from left to right and from bottom to top of the table. This means that hydrogen, carbon, nitrogen, oxygen and sulphur are all small elements. Carbon, nitrogen and oxygen are quite similar in size and so readily link together in a range of biological molecules which almost always carry an outer layer of tiny hydrogen atoms.

1.7 ELECTRON STRUCTURE OF ATOMS

Section 1.4 discussed the structure of atoms which were found to consist of a small, massive nucleus made up of protons and neutrons surrounded by tiny, light electrons.

• **Figure 1.1** Periodic table of the natural elements.

$^{1}_{1}$H 1.01																	$^{4}_{2}$He 4.00
$^{7}_{3}$Li 6.94	$^{9}_{4}$Be 9.01											$^{11}_{5}$B 10.8	$^{12}_{6}$C 12.0	$^{14}_{7}$N 14.0	$^{16}_{8}$O 16.0	$^{19}_{9}$F 19.0	$^{20}_{10}$Ne 20.2
$^{23}_{11}$Na 23.0	$^{24}_{12}$Mg 24.3											$^{27}_{13}$Al 27.0	$^{28}_{14}$Si 28.1	$^{31}_{15}$P 31.0	$^{32}_{16}$S 32.1	$^{35}_{17}$Cl 35.5	$^{40}_{18}$Ar 39.9
$^{39}_{19}$K 39.1	$^{40}_{20}$Ca 40.1	$^{45}_{21}$Sc 45.0	$^{48}_{22}$Ti 47.9	$^{51}_{23}$V 50.9	$^{52}_{24}$Cr 52.0	$^{55}_{25}$Mn 54.9	$^{56}_{26}$Fe 55.8	$^{59}_{27}$Co 58.9	$^{59}_{28}$Ni 58.7	$^{64}_{29}$Cu 63.5	$^{65}_{30}$Zn 65.4	$^{70}_{31}$Ga 69.7	$^{73}_{32}$Ge 72.6	$^{75}_{33}$As 74.9	$^{79}_{34}$Se 79.0	$^{80}_{35}$Br 79.9	$^{84}_{36}$Kr 83.8
$^{85}_{37}$Rb 85.5	$^{88}_{38}$Sr 87.6	$^{89}_{39}$Y 88.9	$^{91}_{40}$Zr 91.2	$^{93}_{41}$Nb 92.9	$^{96}_{42}$Mo 95.9	$^{98}_{43}$Tc *98*	$^{101}_{44}$Ru 101	$^{103}_{45}$Rh 103	$^{106}_{46}$Pd 106	$^{108}_{47}$Ag 108	$^{112}_{48}$Cd 112	$^{115}_{49}$In 115	$^{119}_{50}$Sn 119	$^{122}_{51}$Sb 122	$^{128}_{52}$Te 128	$^{127}_{53}$I 127	$^{131}_{54}$Xe 131
$^{133}_{55}$Cs 133	$^{137}_{56}$Ba 137	$^{139}_{57}$La* 139	$^{178}_{72}$Hf 178	$^{181}_{73}$Ta 181	$^{184}_{74}$W 184	$^{186}_{75}$Re 186	$^{190}_{76}$Os 190	$^{192}_{77}$Ir 192	$^{195}_{78}$Pt 195	$^{197}_{79}$Au 197	$^{201}_{80}$Hg 201	$^{204}_{81}$Tl 204	$^{207}_{82}$Pb 207	$^{209}_{83}$Bi 209	$^{209}_{84}$Po *209*	$^{210}_{85}$At *210*	$^{222}_{86}$Rn *222*
$^{223}_{87}$Fr *223*	$^{226}_{88}$Ra *226*	$^{227}_{88}$Ac** *227*															

$^{1}_{1}$H 1.01 Major biosphere element

$^{51}_{23}$V 50.9 Biosphere trace element

$^{7}_{3}$Li 6.94 Biosphere inactive element

$^{1}_{1}$H 1.01 — Mass number of major isotope; Atomic number; Relative atomic mass

Electrons in an atom are arranged in shells of increasing energy

The way in which the electrons are arranged determines the properties of the atom and thus those of the element. Electrons are organised into a series of energy levels and some simple rules govern their behaviour. The electrons are arranged in levels of increasing energy called **shells**. The first, lowest energy, shell can hold one or two electrons. The second shell can hold between one and eight electrons. The third electron shell can take up to 18 electrons but it usually holds up to eight. An atom seeks to exchange or share electrons with other atoms to achieve a complete outer shell of electrons. This is because a full shell of electrons is very stable. Higher energy levels are more reactive and so only this outer shell of electrons takes part in biochemical or chemical reactions.

Using these ideas, atomic structure diagrams can be drawn for each element to show the electron structure. When the number of electrons in each shell is stated, starting with the number in the lowest energy shell, this is called the **electron structure** of the element.

WORKED EXAMPLE 1.3

Oxygen is present in proteins, carbohydrates and lipids. Give the atomic structure diagram for oxygen and state the electron structure.

ANSWER

From Table 1.4 oxygen has atomic number 8 and mass number 16.
Thus it has 8 electrons and 16 – 8 = 8 neutrons.
The atomic structure diagram is drawn to show the number of protons and neutrons in the nucleus and the electrons in shells, two in the first shell and six in the second as in Figure 1.2.
The electron structure is: O 2.6.

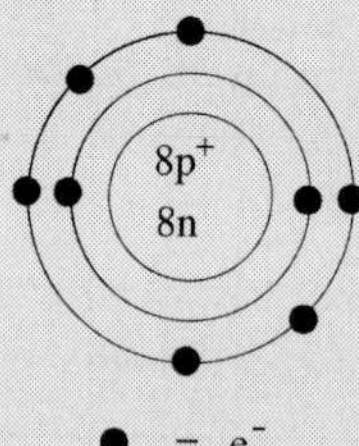

• **Figure 1.2** Atomic structure diagram for oxygen.

WORKED EXAMPLE 1.4

Calcium occurs in the skeletons of many vertebrates. Give the electron structure for the element and construct an atomic structure diagram.

ANSWER

Calcium has atomic number 20 and mass number 40 (Figure 1.1), thus it has 20 electrons and 40 – 20 = 20 neutrons.
The electron structure is: Ca 2.8.8.2.
The atomic structure diagram (Figure 1.3) shows the 20 protons and 20 neutrons in the nucleus together with the four shells of electrons.

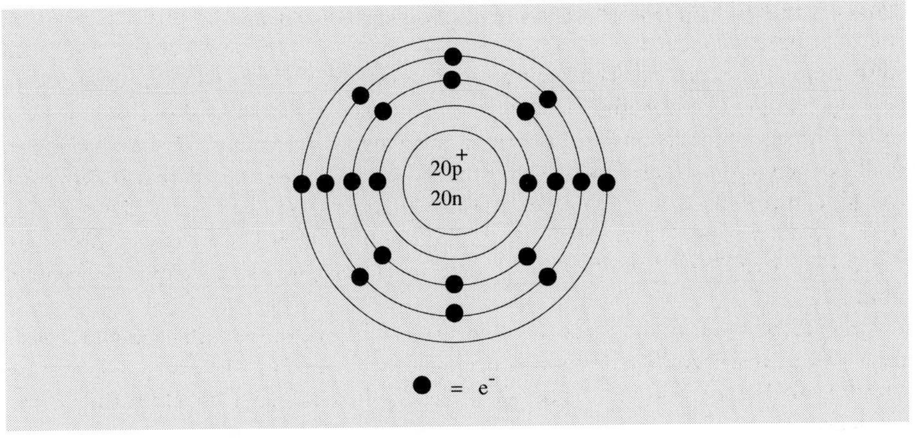

• **Figure 1.3** Atomic structure diagram for calcium.

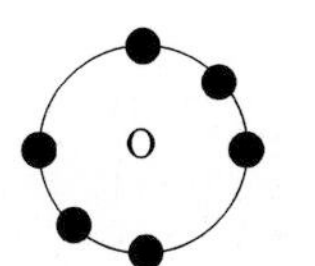

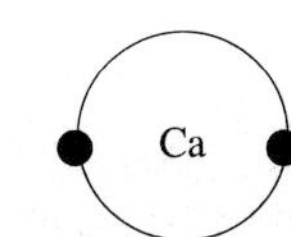

• **Figure 1.4** Simplified atomic structure diagrams for oxygen and calcium.

The full atomic structure diagrams are often simplified to show only the outer shell electrons around the element symbol which represents the nucleus and inner full shells of electrons. In this way the diagrams for oxygen (Figure 1.2) and calcium (Figure 1.3) are reduced to those given in Figure 1.4.

QUESTION 1.3

The element chlorine is implicated in the mechanism of nerve impulses; draw full and simplified atomic structure diagrams for it and give the electron structure.

The atomic structure diagrams just described and the electron structures linked to them can be used to describe a range of interactions between the atoms of different elements (Chapter 2). However, these ideas do not always give results which correspond closely with experimental observations. The approach of a theory called quantum mechanics can be used to explain accurately the behaviour of electrons in atoms. A discussion of the theory is beyond the scope of this book; a full description is provided in *Advanced Inorganic Chemistry* by Cotton *et al.* (1999). Parts of the volume of space around the nucleus have different probabilities of being occupied by electrons. Regions of high probability are called **electron orbitals**. Quantum mechanics predicts a number of orbitals, each one corresponding to a specific energy level for the electron.

An orbital is a volume of space with a high probability of finding an electron

In addition to the existence of orbitals, the theory predicts several features associated with them:

1. There are several different types of orbitals each with a characteristic shape. Three important types are labelled s, p and d. An **s orbital** can be visualised as a sphere with an atomic nucleus at its centre (Figure 1.5). A **p orbital** is shaped like a dumbell with the nucleus situated at the point where the two lobes join (Figure 1.5). A **d orbital** has four lobes, with the nucleus at the centre of the cluster (Figure 1.5).

s orbitals and p orbitals are important in biosphere elements

Table 1.5 Number and type of orbitals in electron shells of an atom

Shell	*Total number of orbitals*	*Orbital type and number*		
		s	*p*	*d*
1	1	1	0	0
2	4	1	3	0
3	9	1	3	5

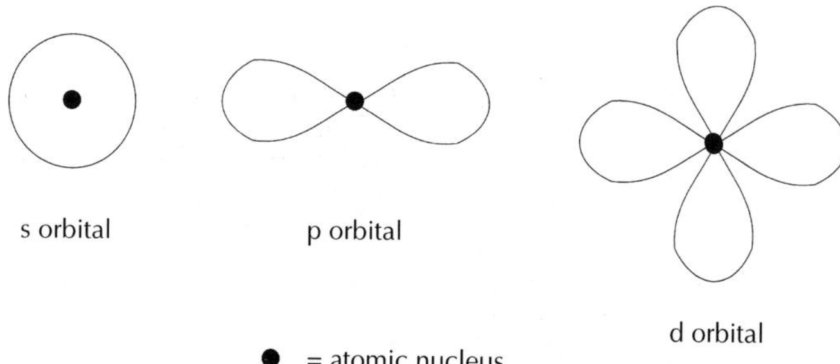

• **Figure 1.5** Atomic orbitals.

An orbital can hold two electrons of opposite spin

2. Each orbital can hold only a maximum of two electrons. This idea is known formally as the Pauli exclusion principle. Electrons are regarded as having a property which can be interpreted as spin, with each electron spinning either 'clockwise' or 'anticlockwise'. When two electrons are present in an orbital, the spins must be 'paired', with one spinning clockwise and the other anticlockwise (Figure 1.6).

Orbitals are arranged in shells within an atom

3. The orbitals are arranged in a series of **shells** which increase in size and thus extend further from the atomic nucleus. The shells are numbered sequentially 1, 2, 3, . . . The energy of the shells increases in the same sequence.
4. The electron shells can hold an increasing number of orbitals in a fixed pattern as shown in Table 1.5. This confirms that there are n^2 orbitals in the nth shell and that the number of each type of orbital increases in the sequence of odd numbers:

• **Figure 1.6** Two spin-paired electrons in an orbital.

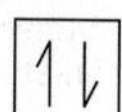

s, p, d, . . .   1, 3, 5, . . .

The three p orbitals are arranged mutually at right angles to one another along the x, y and z axes; they are called the p_x, p_y and p_z orbitals, respectively (Figure 1.7). Within a shell,

• **Figure 1.7** The three p orbitals.

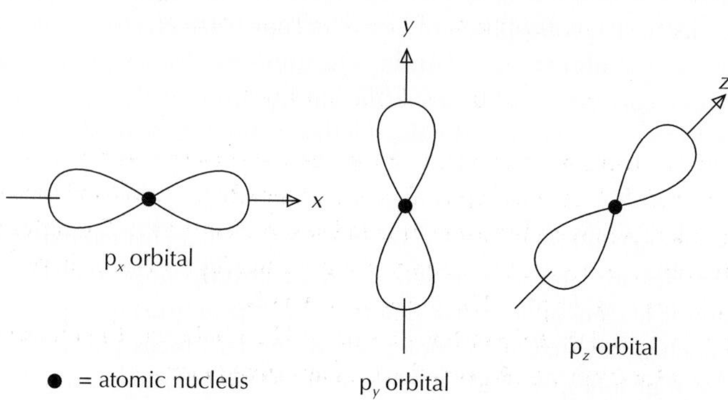

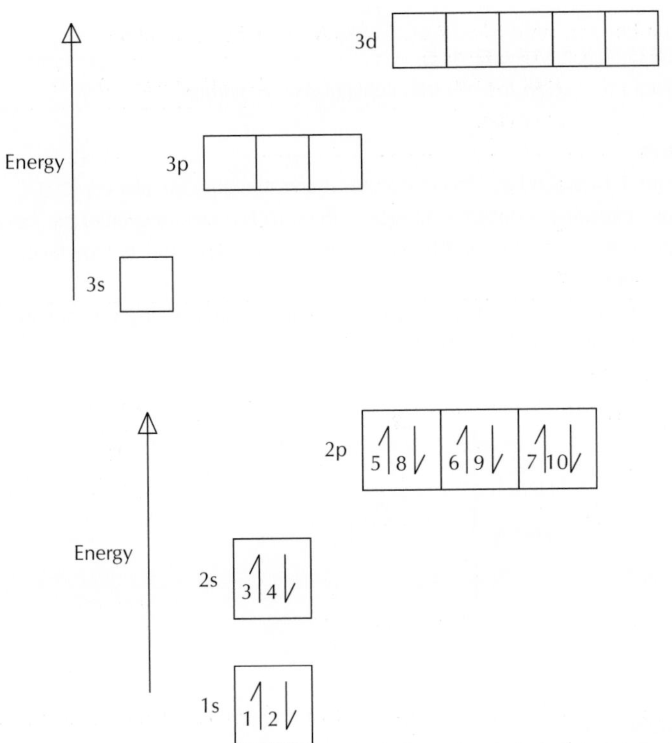

• **Figure 1.8** The arrangement of orbitals in the third shell.

• **Figure 1.9** The sequence of electron occupancy of orbitals in shells 1 and 2.

the orbitals occupy **sub-shells** where the orbitals are of equal energy. Thus the three p orbitals form a sub-shell as do the five d orbitals. In the second shell the four orbitals exist as a lower energy sub-shell with one s orbital and a higher energy sub-shell with three p orbitals of equal energy. The third shell carries nine orbitals in three sub-shells, the lowest energy s sub-shell with a single orbital, a high energy p sub-shell with three orbitals of equal energy and a highest energy d sub-shell with five orbitals of equal energy (Figure 1.8).

A sub-shell contains orbitals of equal energy

When electrons are added to the orbitals within an atom, they enter in a fixed sequence; they occupy the orbitals of lowest available energy first and when a sub-shell of equal energy orbitals is present, they occupy each orbital first before pairing in any one. This effect is known as Hund's rule; it can occur when electrons enter p or d orbitals. The sequence of electron occupancy of the two lowest energy shells is shown in Figure 1.9. This figure can be represented in a shorthand form as:

Electrons always enter orbitals of lowest energy first

$$(1s)^2(2s)^2(2p_x)^2(2p_y)^2(2p_z)^2$$

The number of electrons in an orbital is shown as a superscript number after the orbital symbol. The second shell can hold up to eight electrons when it is full and then the structure is very stable. Elements with between one and seven electrons in the outer second shell will react with other elements by gaining, losing or sharing electrons in order to achieve a full shell or to leave it empty. This is the **octet rule**.

The principles 1 to 4 just described can be used to build up the electron structure, often called the **electron configuration**, of any given element.

WORKED EXAMPLE 1.5

Draw a diagram to show the electron configuration of carbon.

ANSWER

Atoms such as carbon try to gain a share in eight electrons for the outer shell

From Figure 1.1 carbon has atomic number 6 and therefore six electrons.
The first two electrons occupy the 1s orbital, the next two electrons enter the 2s orbital, the next electron enters the $2p_x$ and the last goes into the $2p_y$. The energy level diagram is shown in Figure 1.10.
It is important to note that the sixth electron goes into the $2p_y$ and not the $2p_x$. The shorthand representation is: C $(1s)^2(2s)^2(2p_x)^1(2p_y)^1$.

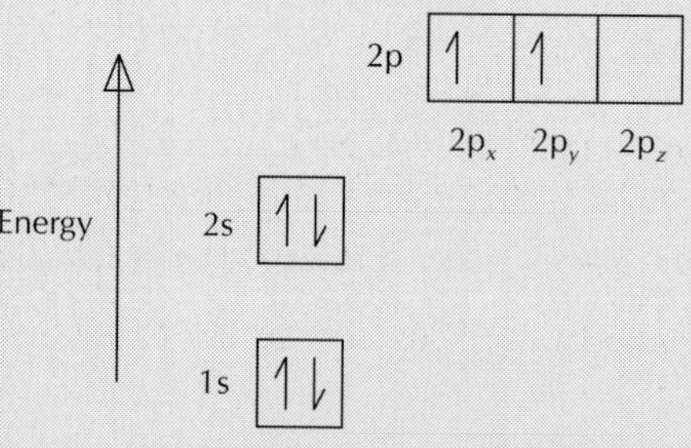

• **Figure 1.10** Energy level diagram showing the arrangement of electrons in the carbon atom.

QUESTION 1.4

Draw a suitable energy level diagram to show the electron configuration of the oxygen atom.

■ 1.8 THE MOLE CONCEPT

A mole of any substance contains the same number of molecules of that substance

A mole is a measure of the amount of a substance present. One mole of any substance contains the same number of atoms or molecules as a mole of any other substance. The number of atoms or molecules in a mole is called the **Avogadro constant** after the scientist who first invented the concept. It is a very large number, equal to $6.022\ 1367 \times 10^{23}$, or 602 213 670 000 000 000 000 000 molecules per mole.

A mole of any substance is the molecular mass of that substance expressed in grams

In terms of mass, a mole of any substance is the molar mass of that substance expressed in grams. The number of moles of any substance in a given mass can be calculated from the formula:

$$\text{Number of moles} = \frac{\text{Mass in grams}}{\text{Molar mass}}$$

■ CALCULATING MOLAR MASSES

The molar mass of a compound is calculated by adding together the atomic masses of the atoms in its molecule. The atomic masses are obtained from tables.

■ MOLARITY

The concentration of a solution is usually measured in moles per litre (mol l^{-1} or mol dm^{-3}). This measure is usually called the **molarity** of the solution. A one molar (1 M) solution contains 1 mole of a solute in 1 litre (1 dm^3) of solution. Measuring concentration in this way enables us to compare directly the number of molecules present in given volumes of different solutions.

■ SUMMARY

All matter, including living organisms, is composed of elements. Only a few of the naturally occurring elements are important in the biosphere. Each element is represented by a symbol and is made up of many tiny atoms while in turn an atom is formed from protons, neutrons and electrons. The number of neutrons in the atoms of an element can vary; the different forms are called isotopes. The arrangement of electrons in an atom is determined by rules which cause the electrons to be placed in orbitals and these are organised into shells of increasing energy.

■ SELECTED FURTHER READING

Hill, G. and Holman, J. (1995) 'Electronic structure', in *Chemistry in Context*, Ch. 6, 4th edn. Nelson, London.

Cotton, F.A. and Wilkinson, G. (1994) 'Electronic structure of atoms', in *Basic Inorganic Chemistry*, Ch. 2, 3rd edn. Wiley, Chichester.

Cotton, F.A., Murillo, C., Wilkinson, G., Bochmann, M. and Grimes, R. (1999) *Advanced Inorganic Chemistry*. Wiley, Chichester.

■ END OF CHAPTER QUESTIONS

Question 1.5 Write symbols for the elements boron, potassium, cobalt, iodine, calcium and molybdenum.

Question 1.6 For each of the following isotopes, $^{1}_{1}H$, $^{2}_{1}H$, $^{14}_{6}C$, $^{31}_{15}P$, $^{37}_{17}Cl$ give:

(a) the relative atomic mass (A_r),
(b) the atomic number,
(c) the number of neutrons,
(d) the number of electrons.

Question 1.7 Draw simple electron structure diagrams to show the electron arrangement in the elements:

(a) sodium,
(b) boron,
(c) phosphorus.

Question 1.8 For the elements:

(a) hydrogen,
(b) nitrogen,
(c) sodium,

(i) construct energy level diagrams to represent the sequence of electron orbitals and shells in the atom,
(ii) write the shorthand notation for the electron configuration in each case.

2 COVALENT BONDING AND MOLECULES

2.1 INTRODUCTION

A living organism derives most of its character from the enormous range of molecules contained within it. These help to determine the structure of the body, the function of enzymes, the clotting of blood, cell respiration and innumerable other features. It is useful for us to understand the structure of molecules and to consider their properties in order to interpret the role they perform in the organism.

2.2 INTERACTIONS BETWEEN ATOMS

Atoms share electrons to form covalent bonds

Proteins, sugars and other biologically important molecules are collections of atoms held together by a force of attraction called **covalent bonding**. **Covalent bonds** are formed between atoms that can share electrons to achieve a full shell of electrons. Such a full shell is a very stable structure. When two atoms move towards one another, the outer shell electrons in each atom rearrange themselves to reduce their potential energy. The energy falls to a minimum when they are a specific distance apart. This distance, measured between the two atomic nuclei, is called the **bond length**. The fall in energy that has occurred is the **bond energy** (Figure 2.1).

2.3 COVALENT BONDS FORMED BY SHARING OUTER ELECTRONS

Only outer shell, or valence shell, electrons are shared in bonding

Electrons that form a bond are localised in the area between the atoms

The simplest way to describe the formation of a covalent bond is to consider the way in which the outer shells of electrons in the atoms interact as the atoms approach one another. The outer shell of electrons is often called the **valence shell**. So this description of covalent bonding is called the **valence-bond theory**. The outer electrons interact by sharing themselves between the atoms to form overlapping full shells. For example, the element hydrogen has one electron and needs to share a second electron in order to obtain a full shell of two electrons. The biologically significant elements carbon, nitrogen, oxygen, phosphorus, sulphur and chlorine all require a total of eight shared electrons to form a full outer shell. The bonding process can be illustrated by the formation of the hydrogen molecule. Two hydrogen atoms each with one electron in the first shell (electron structure H 1) (Table 2.1) join together so that the shells overlap and each

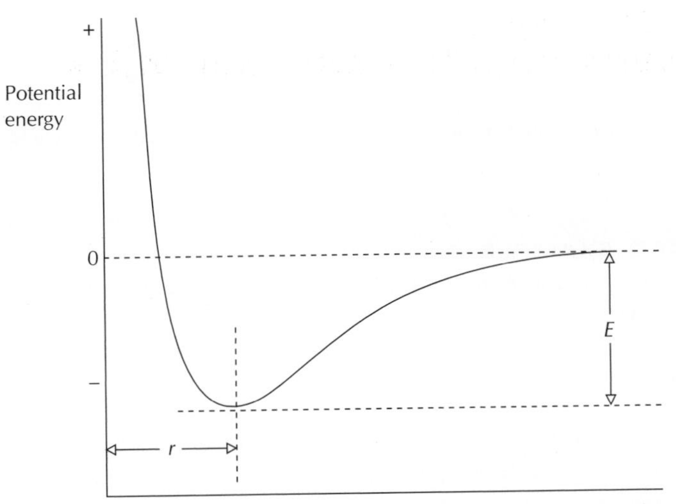

• **Figure 2.1** Potential energy change which occurs when two atoms are brought together to form a chemical bond. The fall in energy *E* corresponds to the energy of the chemical bond. The distance *r* between the nuclei of the two atoms represents the potential energy minimum and is the bond length.

Table 2.1 Names, symbols, electron structure and valencies (oxidation numbers) for elements of biological importance

Name	*Symbol*	*Electron structure*	*Valency*
Hydrogen	H	1	1
Carbon	C	2.4	4
Nitrogen	N	2.5	3 or 5
Oxygen	O	2.6	2
Sodium	Na	2.8.1	1
Magnesium	Mg	2.8.2	2
Phosphorus	P	2.8.5	3 or 5
Sulphur	S	2.8.6	2
Chlorine	Cl	2.8.7	1
Potassium	K	2.8.8.1	1
Calcium	Ca	2.8.8.2	2

shell gains a share in two electrons (Figure 2.2) and is full. The two atoms are linked strongly together to form a hydrogen molecule. The two shared electrons are concentrated in the volume of space between the two hydrogen nuclei to form the covalent bond. The molecule is often shown as the two symbols for the element linked by a line to indicate the bond, H—H; or more simply as H_2. The same ideas can be used to describe the formation of covalent bonds in simple biological molecules.

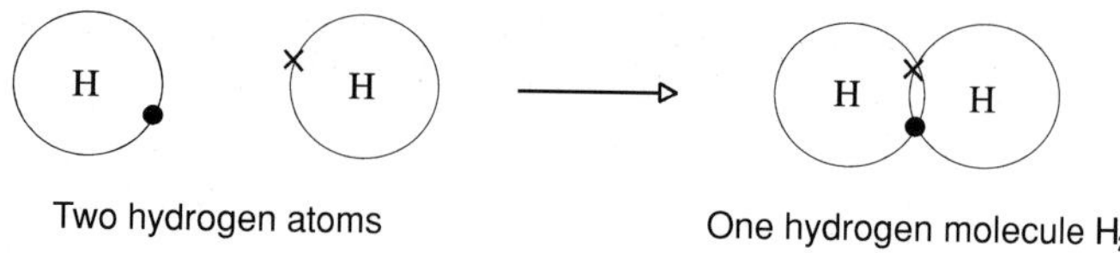

• **Figure 2.2** Formation of a covalent bond in the hydrogen molecule.

WORKED EXAMPLE 2.1

The vital solvent medium in the living cell is water. Describe the covalent bond in the water molecule, H_2O.

ANSWER

Write the electron structures for hydrogen and oxygen (Table 2.1): H 1 O 2.6

Oxygen has six electrons in the outer shell and therefore needs to share in two extra electrons to gain a full shell of eight.

Two hydrogen atoms are required to provide these two electrons.

Next, draw diagrams to show the outer shell electrons of one oxygen atom and two hydrogen atoms and then allow the outer shells of electrons to overlap (Figure 2.3).

Diagrams showing overlap of outer shell electrons are used to explain covalent bonding

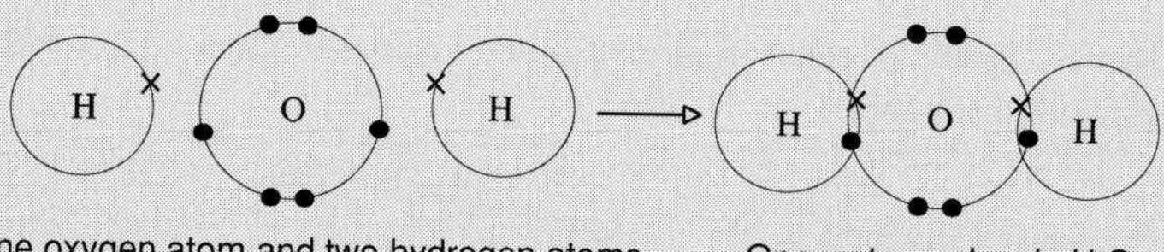

One oxygen atom and two hydrogen atoms One water molecule H_2O

• **Figure 2.3** Formation of two covalent bonds in the water molecule.

WORKED EXAMPLE 2.2

Methane, CH_4, the simplest organic compound, is the starting point for understanding the complex range of biological molecules. Use diagrams to construct the covalent bonds in methane.

```
   H
   |
H—C—H
   |
   H
```

The structural formula of methane

ANSWER

Write the electron structures for carbon and hydrogen (Table 2.1): C 2.4 H 1

Carbon has four electrons in the outer shell. To form a full outer shell of eight and thus a stable structure it needs to gain a share in four electrons.

Four hydrogen atoms are required to provide these electrons. Thus the formula is confirmed as CH_4.

Now draw a carbon atom and four hydrogen atoms showing only the outer shell electrons. Bring the atoms together allowing the outer shells to overlap in a diagram of the molecule (Figure 2.4).

The water and methane molecules can be shown in the same simplified form as the hydrogen molecule, with lines to indicate the bonds.

QUESTION 2.1

Amino acids are derived from the ammonia molecule, NH_3. Draw diagrams to show how the covalent bonds in the molecule are formed and write a simplified form of the structure to show the bonds.

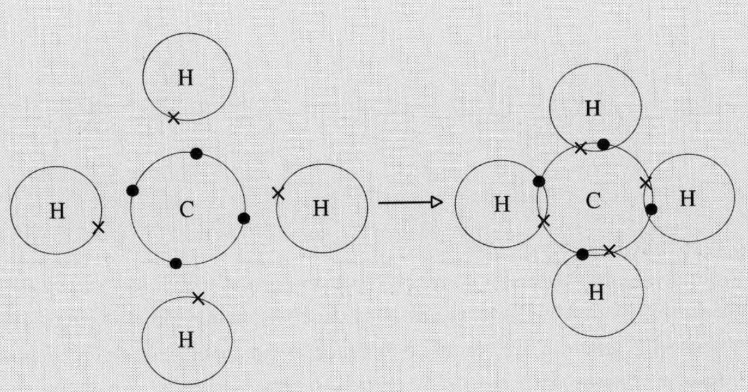

• **Figure 2.4** Formation of the methane molecule with four covalent bonds.

QUESTION 2.2

Sulphur bacteria use the compound hydrogen sulphide, H_2S, to provide hydrogen for the reduction of carbon dioxide to form sugars. Draw diagrams to explain how covalent bonds are formed in the hydrogen sulphide molecule.

Outer electron shells in atoms can share more than one pair of electrons in forming covalent bonds

The form of the methane and water molecules was achieved by atoms sharing electrons in pairs to give covalent bonds. It is possible for a pair of atoms to share more than one pair of electrons to form two or three covalent bonds. Oxygen, O_2, which is vital for the process of respiration, is an important example of a molecule which shares two pairs of electrons between the atoms. The electron structure of the oxygen atom is O 2.6 (Table 2.1). Thus it requires a share in two extra electrons to obtain a full outer shell. These can be supplied by a second oxygen atom which in turn shares two electrons from the first atom. This is conveniently shown by a diagram of the oxygen atoms and the oxygen molecule formed (Figure 2.5). The molecule is represented in a simple form by linking the element symbols by two lines, O=O.

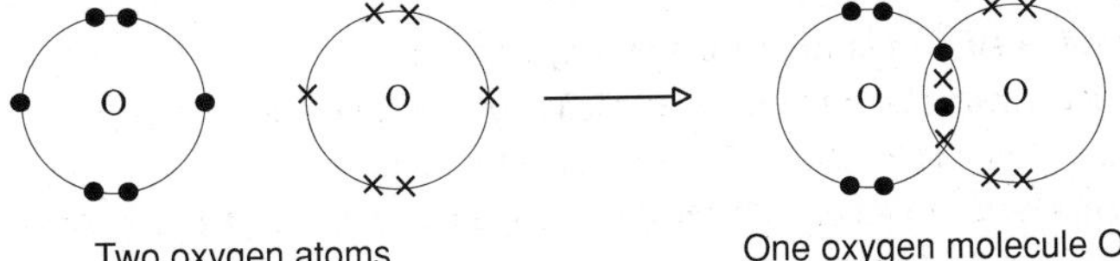

• **Figure 2.5** Double covalent bond formation in the oxygen molecule.

WORKED EXAMPLE 2.3

The gaseous respiration product carbon dioxide, CO_2, is a covalent molecule. Draw diagrams to show the formation of the carbon–oxygen bonds.

ANSWER

First write the electron structures of carbon and oxygen (Table 2.1): C 2.4 O 2.6

Then draw diagrams to show the outer shell electrons for one carbon and two oxygen atoms.

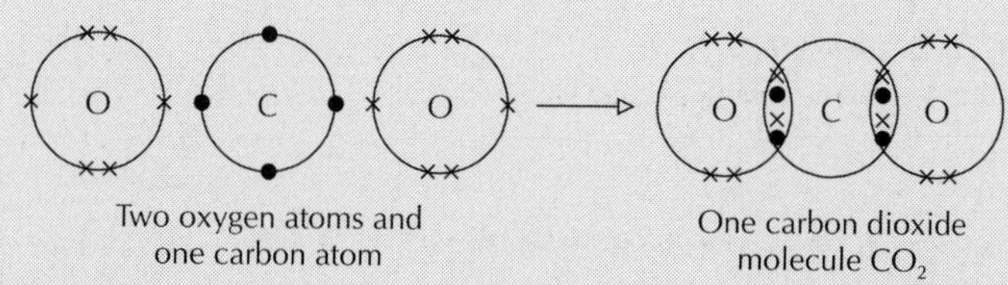

• **Figure 2.6** Double covalent bond formation in the carbon dioxide molecule.

Carbon, with four outer shell electrons, requires a share in four more to give a full shell of eight. These are provided by two oxygens each giving two electrons. The oxygen atoms gain, at the same time, a share in two electrons to increase the six electrons to a full shell of eight. Finally, allow the three atoms to come together with overlap of the outer shells of the electrons (Figure 2.6). The simple representation of the carbon dioxide molecule is O=C=O.

QUESTION 2.3

Nitrogen gas, N_2, is the main component of the earth's atmosphere. It provides the ultimate source of nitrogen for amino acids and proteins. Draw diagrams to show how multiple covalent bonding occurs in the nitrogen molecule.

■ 2.4 FORMULAE OF COMPOUNDS

The number of outer shell electrons shared by an element in bonding is the valency

Simple rules are used to determine the formula of a compound from the valency

In discussing the formation of covalent bonds, it has been useful to write the formulae for molecules using element symbols. It was clear that the number of electrons needed by each element to complete its outer shell determined the ratio of the element within the given molecule. When we consider the same element in different molecules, it always requires a share in the same number of extra electrons. Thus hydrogen always shares one electron, while carbon requires four electrons and oxygen needs two electrons. The number shared is called the **valency** or combining power of the element. More formally, it is known as the **oxidation number**. These valencies can be used conveniently to determine the formulae of compounds in a simple way, without the need to consider outer electron shell structures.

In order to write the formula of the water molecule, the symbols and valencies for the elements are used.

WORKED EXAMPLE 2.4

Write the formula for the water molecule.

ANSWER

Water contains the elements hydrogen and oxygen.

Write the symbols:	H	O
Write the valency for each element (Table 2.1):	1	2
Change the valency of each element to the other one, writing the numbers below the line and after the element symbols:	H_2	O_1
Close up the numbers and letters, ignore the number '1' to give the formula:	H_2O	

WORKED EXAMPLE 2.5
Give the formula for the methane molecule.

ANSWER
Methane contains the elements carbon and hydrogen.

Write the symbols:	C	H
Write the valency for each element (Table 2.1):	4	1
Change over the valencies from one element to the other and write below the line and after the element symbol:	C_1	H_4
Close up and ignore the number '1' to obtain the formula:	CH_4	

WORKED EXAMPLE 2.6
Determine the formula of the carbon dioxide molecule.

ANSWER
Carbon dioxide contains the elements carbon and oxygen.

Write the symbols:	C	O
Write the valency for each element:	4	2
Exchange the valencies from one element to the other, writing each below the line after the symbols:	C_2	O_4
In this case, the numbers can be simplified by dividing by two before closing up to give the formula:	CO_2	

When the numbers obtained initially can be simplified by dividing by a common value, then this is usually carried out.

QUESTION 2.4
Use valencies to write the formula of ammonia, which contains the elements nitrogen and hydrogen.

QUESTION 2.5
Determine the formula for hydrogen sulphide, which contains hydrogen and sulphur, by using element valencies.

Groups of atoms as well as single atoms can have a valency

When a compound contains more than two elements the same ideas can be used to determine the formula. The process is often made simpler because groups of atoms usually stay together in a number of different compounds. Thus the carbonate group containing a carbon atom and three oxygen atoms, CO_3, occurs in many compounds such as calcium carbonate, $CaCO_3$, sodium carbonate, Na_2CO_3, and magnesium carbonate, $MgCO_3$. In each case, the valency of two can be assigned to the carbonate group and the formula determined on this basis. A selection of these groups and valencies is collected in Table 2.2.

Table 2.2 Names, symbols and valencies for some biologically significant groups

Group	*Formula*	*Valency*
Amide	$CONH_2$	1
Amine	NH_2	1
Ammonium	NH_4	1
Carbonate	CO_3	2
Carboxylic acid	COOH	1
Bicarbonate (hydrogencarbonate)	HCO_3	1
Hydroxide	OH	1
Alcohol	OH	1
Ketone	CO	2
Sulphate	SO_4	2
Phosphate	PO_4	3

WORKED EXAMPLE 2.7

Calcium carbonate is a major component of the exoskeleton of phytoplankton species. Use valencies to determine the formula of calcium carbonate.

ANSWER

Calcium carbonate contains the element calcium and the carbonate group (Table 2.2).

Write the element and group symbols:	Ca	CO_3
Write the valency for each (Tables 2.1 and 2.2):	2	2
Change valency for element to the group and vice versa. Write below the line and after the symbol:	Ca_2	$CO_{3\,2}$
Simplify the numbers by dividing by two, the resulting number '1's are not written. Close up symbols to give the formula:	$CaCO_3$	

WORKED EXAMPLE 2.8

Ammonium carbonate is formed from the products of protein decay, ammonia and carbon dioxide, dissolved in water. What is the formula for ammonium carbonate?

ANSWER

Ammonium carbonate contains the groups ammonium and carbonate (Table 2.2).

Write the formulae for the groups with the valencies for each beneath (Table 2.2):	NH_4 1	CO_3 2
Interchange the valencies between the groups, write below the line and after the formula:	$NH_{4\,2}$	$CO_{3\,1}$
Close up the group formulae omitting the number '1' and placing a bracket around the NH_4 to avoid reading the four and two as forty-two and to show that there are two NH_4 groups, and not just two H_4 groups:	$(NH_4)_2CO_3$	

WORKED EXAMPLE 2.9

Calcium phosphate is an important mineral in the structure of bone. Find the formula of calcium phosphate.

ANSWER

Calcium phosphate contains the element calcium and the phosphate group (Table 2.2).

Write the symbol for calcium and the formula for the phosphate group with valencies underneath (Table 2.2): Ca PO_4 / 2 3

Exchange valencies with one another and write after the symbol or formula and below the line: Ca_3 $PO_{4\,2}$

Place the group formula, PO_4, in a bracket and close up to give the formula of the compound: $Ca_3(PO_4)_2$

QUESTION 2.6

Determine the formula of potassium sulphate which is a source of soluble potassium for the biosphere. The compound contains the sulphate group (Table 2.2).

QUESTION 2.7

Use valencies to find the formula of the fertiliser ammonium phosphate which contains the ammonium and phosphate groups (Table 2.2).

■ 2.5 COVALENT BONDS FORMED BY COMBINING ATOMIC ORBITALS

The formation of covalent bonds by the sharing of electrons between atoms has been discussed in section 2.3. It was seen that this description of molecules was quite simple. Only a few electrons were involved and the results were easy to obtain and understand. However, it is found that a number of molecules do not have the properties that would be expected from a valence-bond description. For example, the oxygen molecule is very reactive, as is clear from its role in respiration. This reactivity can be ascribed to the presence of two electrons in the molecule which are unshared. They do not form a pair as is expected by the overlap of outer-shell electrons to form electron pairs. This means that a more appropriate theory is needed to explain the behaviour of oxygen and many other molecules that possess unexpected properties.

Explaining covalent bond formation by sharing of valence shell electrons is not always satisfactory

In discussing a new approach to the electron structure of molecules, it is important that the ideas of valence-bond theory are not abandoned. They can continue to be used whenever they provide a suitable description of a molecule. It is normal to pick out the best theory for use with any example.

The idea of atomic orbitals was considered in section 1.7; it can be extended to describe orbitals within molecules. Several of the same rules apply to **molecular orbitals** as to atomic orbitals:

Molecular orbitals are formed by combination of atomic orbitals

- molecular orbitals are occupied by electrons in order of increasing energy,
- each molecular orbital can hold up to two electrons,
- when two or more molecular orbitals have the same energy, electrons enter each orbital singly before pairing in any one.

Molecular orbitals are filled with electrons using the same rules as for atomic orbitals

The formation of molecular orbitals can be illustrated by the interactions that take place when two isolated atoms with atomic orbitals approach one another until the orbitals

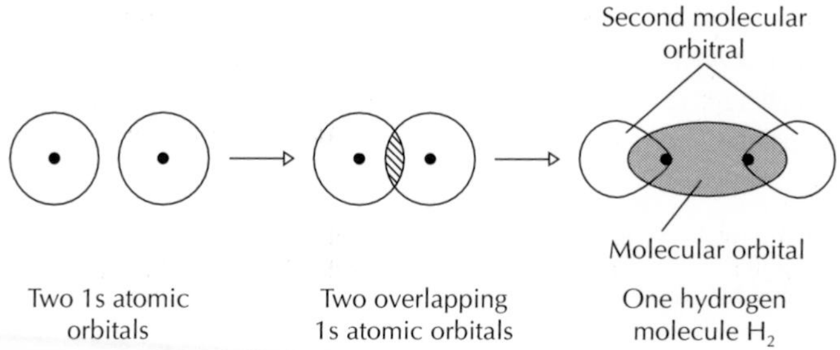

• **Figure 2.7** Overlap of two hydrogen atomic orbitals to form two hydrogen molecular orbitals and a covalent bond.

overlap. In the case of the hydrogen molecule, two hydrogen atoms each with a 1s atomic orbital move together and overlap to produce two new molecular orbitals (Figure 2.7). Although the two atomic orbitals are the same, the two molecular orbitals are quite different. They have different shapes; one is concentrated mainly in the space between the two hydrogen nuclei. The second consists of two parts or **lobes** which are distributed away from the space between the nuclei. The first orbital is of lower energy than either of the original atomic orbitals while the second is of higher energy than the atomic orbitals. The average energy of the two molecular orbitals is the same as the average energy of the two atomic orbitals.

Two electrons are available to occupy the molecular orbitals, one electron being supplied by each hydrogen atom. The electrons enter the molecular orbital of lowest energy, form a pair and fill it. This can be shown on a **molecular orbital energy level diagram** (Figure 2.8).

Placing the two electrons in the low energy molecular orbital concentrates them in the region of space between the two atomic nuclei. It causes a covalent bond to be formed. This orbital is called a **bonding molecular orbital**. The second, higher energy, molecular orbital, if it held electrons, would localise them away from the space between the nuclei. The nuclei would be unshielded from one another and would repel, and no bond would be formed. This is called an **antibonding molecular orbital**.

Molecular orbitals can be of low energy, bonding; or high energy, antibonding

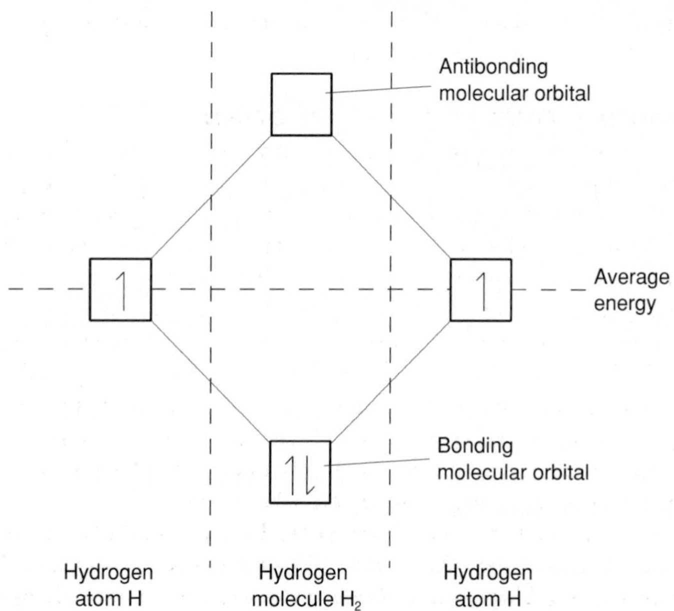

• **Figure 2.8** Molecular orbital energy level diagram for the formation of a hydrogen molecule from two hydrogen atoms.

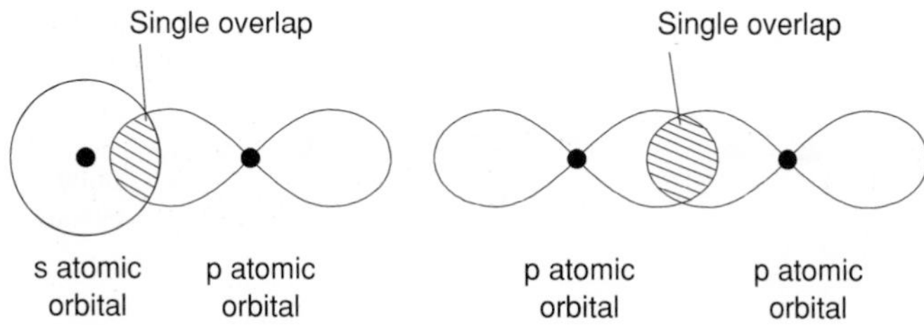

• **Figure 2.9** Single overlap of s and p or p and p atomic orbitals in covalent bond formation.

The ideas used here to describe the formation of a covalent bond in a molecule using molecular orbitals formed from atomic orbitals is called the **molecular orbital theory**. It can be seen that both valence bond and molecular orbital theory can give the same results, that is the formation of a covalent bond by concentration of electron density between the centres of two atoms. But the two theories have arrived at the result by different methods.

■ 2.6 SINGLE OVERLAP, THE SIGMA BOND

For the hydrogen molecule, molecular orbitals are formed by the overlap of two spherical 1s atomic orbitals. From Figure 2.7 it can be seen that the two atomic orbitals have approached each other along a line joining the centres of the two atoms to give a single region of overlap. An s orbital can overlap with a p_x orbital in the same way to give a single region of overlap (Figure 2.9) and thus form two molecular orbitals. Two p_x orbitals can interact by the same process to form, once more, a single region of overlap (Figure 2.9).

The formation of bonding and antibonding molecular orbitals by single overlap is indicated by placing the Greek letter sigma, σ, in front of the orbital name. In Figure 2.8, the two molecular orbitals are then written as a **σ-bonding molecular orbital** and a **σ-antibonding molecular orbital**. The word bonding is usually left out and the term antibonding is replaced by an asterisk, *. Thus the two molecular orbitals are simplified to a **σ molecular orbital** and a **σ* molecular orbital**. In general, any covalent bond formed by single overlap is called a **σ bond**.

Single overlap of atomic orbitals gives a single σ bond

■ 2.7 DOUBLE OVERLAP, THE PI BOND

The three p orbitals are arranged along the *x*, *y* and *z* axes (section 1.7). Single overlap occurs when two p_x atomic orbitals approach each other along the *x*-axis, which is the line joining the centres of the two atoms. When two p_y or two p_z orbitals approach each other along the *x*-axis a different type of overlap takes place. The lobes of the p orbitals overlap sideways in two regions rather than one. The overlap occurs above and below the line joining the two atomic centres (Figure 2.10). The overlap of two p atomic orbitals leads to the formation of two molecular orbitals, a low-energy bonding molecular orbital and a high-energy antibonding molecular orbital. The bonding molecular orbital consists of two lobes, one above and one below the centre line of the new molecule. The antibonding molecular orbital is composed of four parts which are directed away from the region between the atoms (Figure 2.11). It is important to bear in mind that double overlap can only occur after single overlap has taken place.

Double overlap in the formation of two molecular orbitals is indicated by placing the Greek letter pi, π, before the orbital name. Thus the molecular orbitals in Figure 2.11 are written as a **π-bonding molecular orbital** and a **π-antibonding molecular**

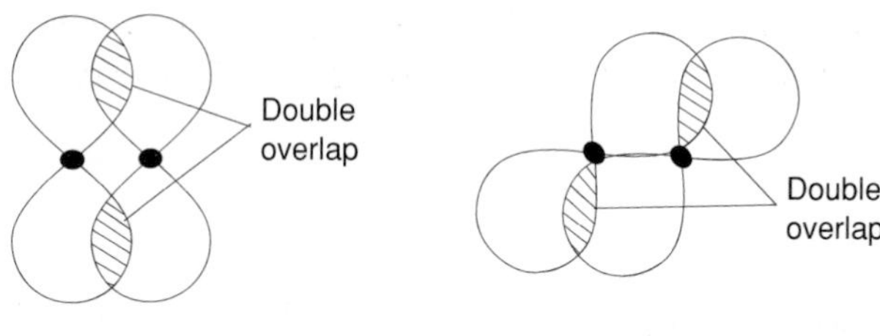

• **Figure 2.10** Sideways interaction of p_z and p_z or p_y and p_y atomic orbitals to give double overlap.

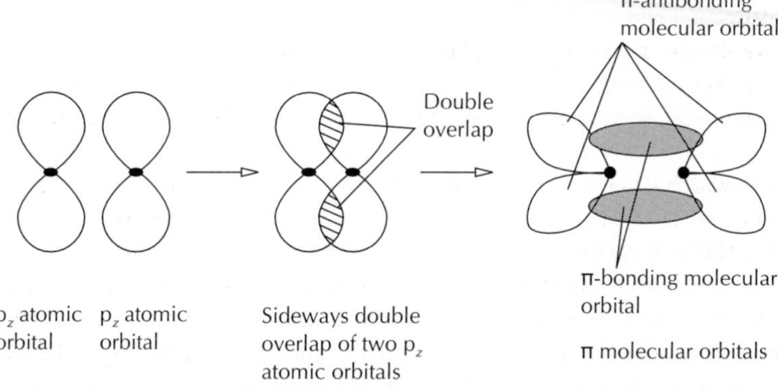

• **Figure 2.11** Sideways overlap of two p atomic orbitals to form two π molecular orbitals and a covalent bond.

orbital or more briefly as a **π molecular orbital** and a **π* molecular orbital**. A covalent bond formed by double overlap is called a **π bond**. Thus it can be stated that a σ bond is always formed before a π bond.

Double overlap of atomic orbitals gives a π bond

Molecular orbitals are filled by electrons in order of increasing energy. This order is:

$$\sigma < \pi < \pi^* < \sigma^*$$

When two molecular orbitals have the same energy then electrons will enter each orbital singly before forming pairs in any one.

■ 2.8 MOLECULES WITH σ AND π BONDS

The hydrogen molecule, H_2, contains a single σ bond, formed by overlap of two s atomic orbitals (section 2.5). In the water molecule, two p atomic orbitals in oxygen each overlap with an s atomic orbital in hydrogen to form two single σ bonds (Figure 2.12).

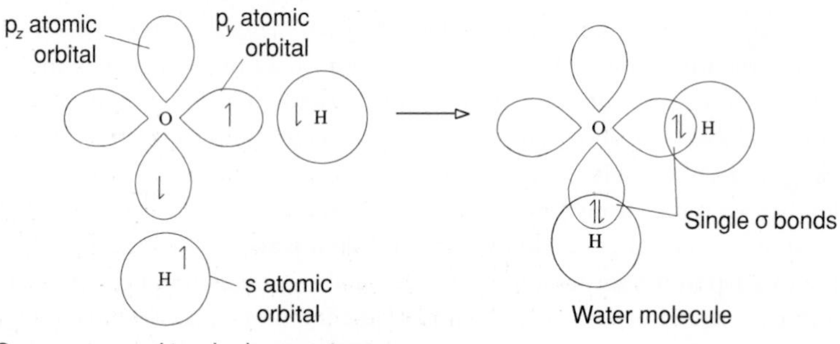

• **Figure 2.12** Overlap of atomic orbitals in hydrogen and oxygen atoms to form covalent σ bonds in the water molecule.

The oxygen molecule, O_2, contains both σ and π bonds. The formation of these bonds can be understood best by constructing the **energy level diagram**.

An energy level diagram can be used to show the formation of σ and π bonds

WORKED EXAMPLE 2.10

Use a molecular orbital diagram to show the bond formation in the oxygen molecule, O_2.

ANSWER

Write the electron configuration of the oxygen atom (Chapter 1).

O $(1s)^2(2s)^2(2p_x)^2(2p_y)^1(2p_z)^1$

Draw the molecular orbital energy level diagram for the two oxygen atoms and the oxygen molecule.

Place the available electrons in the molecular orbitals, filling the lowest energy orbitals first and putting the electrons singly into orbitals of the same energy before pairing in any one (Figure 2.13).

From Figure 2.13 it can be seen that the eight available 2p electrons occupy a σ-bonding molecular orbital, two π-bonding molecular orbitals and two π-antibonding molecular orbitals (one electron in each). Thus there is a total of six electrons in bonding molecular orbitals. The antibonding molecular orbitals contain a total of two electrons.

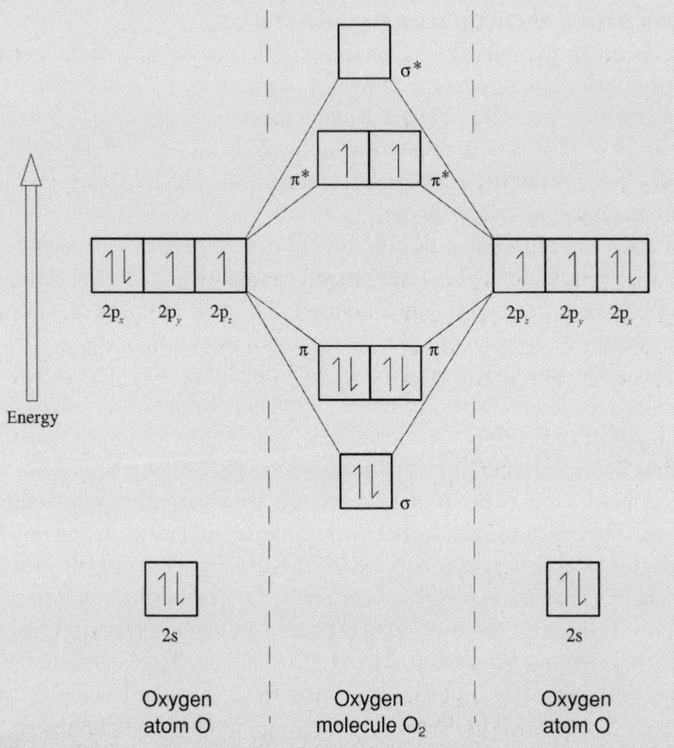

• **Figure 2.13** Molecular orbital energy level diagram to show the formation of the oxygen molecule O_2 from two oxygen atoms.

When electrons are present in bonding molecular orbitals, they cause attraction between the atoms. But electrons in antibonding molecular orbitals cause repulsion between the atoms. Thus to determine the number of bonds in the oxygen molecule, the number of electrons in antibonding orbitals is subtracted from the number in bonding orbitals. Six electrons minus two electrons gives four electrons. Each bond requires two electrons so the number of bonds in oxygen is four divided by two giving two bonds. This can be summarised as follows:

Electrons in bonding molecular orbitals	$= 6$
Electrons in antibonding molecular orbitals	$= 2$
Bonding – antibonding electrons	$6 - 2 = 4$
Number of bonds (two electrons in each bond)	$4 \div 2 = 2$

The single electrons in the π* orbitals give the oxygen molecule its high reactivity.

QUESTION 2.8

Construct the molecular orbital energy level diagram for the nitrogen molecule, N_2, to show the σ and π bonds.

In forming covalent bonds, hydrogen uses only a 1s orbital to form a single σ bond with other elements. The elements carbon, nitrogen and oxygen use 2s and 2p atomic orbitals to form single σ bonds and double or triple π bonds. The heavier elements phosphorus and chlorine use 3s and 3p atomic orbitals to form mainly σ bonds, although phosphorus forms double π bonds, especially with oxygen.

■ 2.9 HYBRID MOLECULAR ORBITALS

Explaining the bonds in molecules by the use of molecular orbitals has helped us to understand the properties of compounds. However, a number of characteristics of molecules still remain hard to interpret. Thus the angle formed by H—O—H bonds in water is known to be 105°. The description of the molecule in section 2.8 suggests a bond angle of 90°. A development of molecular orbital theory called **hybridisation** can be used to explain this anomaly and others like it.

Hybridisation can be justified by considering the overlap of hydrogen 1s and oxygen 2p orbitals shown in Figure 2.12. Only a small part of each 2p orbital takes part in overlap, one lobe in each orbital remaining unused. Since the strength of a bond is proportional to the degree of overlap, this picture suggests a weak bond. The overlap is increased and the bond made stronger by modifying, or hybridising, the orbitals on oxygen. The single 2s orbital combines with the three 2p orbitals to give four new hybrid orbitals called **sp^3 hybrid orbitals**.

One 2s and three 2p orbitals can combine to form four equal sp^3 hybrid orbitals, leading to strong covalent bonds

These four hybrid orbitals are equal and each consists of one large lobe and one very small lobe (Figure 2.14). The sp^3 hybrid orbitals are arranged tetrahedrally around the oxygen atom. They each use the large lobe for overlap and form strong bonds. Thus it is energetically favourable for oxygen to use hybrid orbitals in forming the water molecule. The six available electrons in oxygen occupy each sp^3 hybrid singly before pairing in two of the orbitals. For water, the overlap of orbitals is shown in Figure 2.15 (the small lobe on each hybrid is left out for clarity). The H—O—H bond angle predicted by this theory is the tetrahedral angle 109°, a little larger than the 105° found in the actual molecule. The difference is explained by the strong repulsion between the two pairs of unshared electrons which pushes them apart and reduces the H—O—H angle to 105°.

• **Figure 2.14** An sp^3 hybrid orbital.

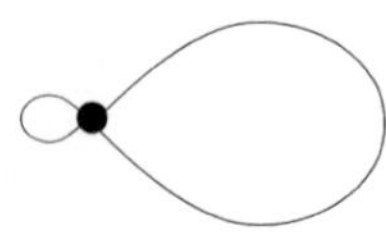

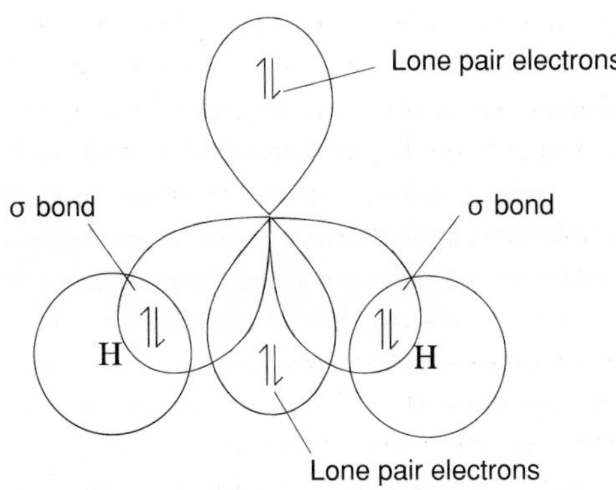

• **Figure 2.15** Overlap of two hydrogen atomic s orbitals with two oxygen sp^3 hybrid orbitals to form two σ bonds. Two sp^3 hybrid orbitals each contain a pair of unshared electrons and form lone pairs.

The unshared pairs of electrons in two of the sp^3 hybrid orbitals occupy a significant region of space. Each protrudes out from the oxygen atom and is called a **lone pair** of electrons. The formation of hybrid orbitals is important in describing the bonds formed by carbon and nitrogen as well as oxygen. Such lone pairs on oxygen or nitrogen atoms are important in determining the properties of proteins and the solvent power of water.

Hybrid orbitals not used to form covalent bonds contain unshared electron pairs called lone pairs

WORKED EXAMPLE 2.11

Draw hybrid orbitals to show the formation of four σ bonds in the methane molecule, CH_4.

ANSWER

Write the electron structure of the carbon atom. C $(1s)^2(2s)^2(2p_x)^1(2p_y)^1$

The 2s, $2p_x$, $2p_y$ and the unoccupied $2p_z$ orbitals form four equal sp^3 hybrid orbitals.

The four available electrons on carbon (the 1s electrons remain unused) occupy each hybrid orbital singly.

Four hydrogen atoms each provide a 1s orbital with a single electron for overlap with the sp^3 hybrids to give four equal σ bonds.

Draw the diagram to show the orbitals and the bonds (Figure 2.16).

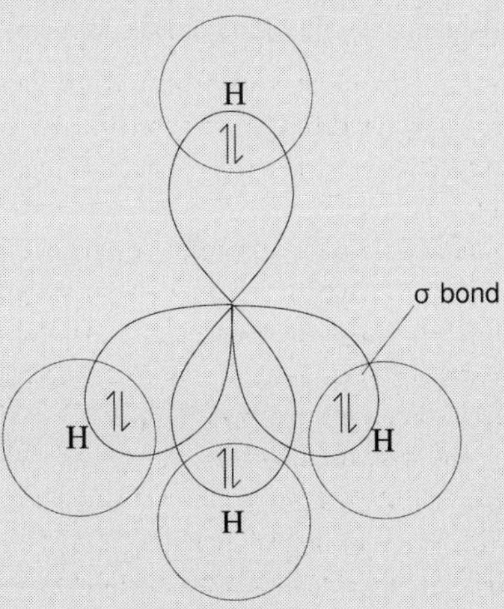

• **Figure 2.16** Overlap of four hydrogen atomic s orbitals with four carbon sp^3 hybrid orbitals to form four σ bonds.

■ SUMMARY

When two atoms share between them a pair of electrons, a covalent bond is formed. Atoms of elements share electrons in order to achieve a full outer shell which is stable. Hydrogen requires a share in two electrons while carbon, nitrogen and oxygen each require a share in eight. A full outer shell can be formed by sharing two or three pairs of electrons in some cases. The number of electrons an element can share is the valency. Valencies are used to determine the formulae of compounds. The formation of covalent bonds can also be explained by the overlap of atomic orbitals to form molecular orbitals. Energy level diagrams are drawn to show the changes in energy which occur on bond formation. Single overlap of atomic orbitals gives σ bonds while double overlap gives π bonds. Bonding and antibonding molecular orbitals are formed by overlap of atomic orbitals. The combination of s and p atomic orbitals gives rise to hybrid orbitals in a number of molecules. Hybrid orbitals form strong bonds and lead to the identification of lone pair electrons.

■ END OF CHAPTER QUESTIONS

Question 2.9 The gas phosphine, PH_3, is a product of anaerobic degradation.

(i) Draw diagrams, using only the outer electron shells of the atoms, to show the formation of covalent bonds in the molecule.

(ii) Write a simple form of the molecule to show the bonds.

Question 2.10 Hydrochloric acid in the stomach is derived from hydrogen chloride, HCl.

(i) Draw diagrams to illustrate the outer shell electrons in the hydrogen and chlorine atoms.

(ii) Combine these diagrams to explain the formation of the covalent bond in hydrogen chloride.

Question 2.11 The salt magnesium sulphate is a source of soluble magnesium for many organisms.

Use valencies to write the formula for the salt.

Question 2.12 Limestone dissolves in rain water to give calcium hydrogencarbonate.

Determine the formula of the compound by using valencies.

Question 2.13 The ammonia molecule, NH_3, contains three σ bonds and a lone pair of electrons.

Use a diagram showing hybrid orbitals to represent the bonds in ammonia.

FORCES WITHIN AND BETWEEN MOLECULES

3

■ 3.1 INTRODUCTION

Several types of interaction can occur between molecules or parts of molecules. A single protein molecule will have within it examples of these interactions. The molecule will consist of a long chain of carbon, nitrogen and oxygen atoms, linked by covalent bonds. Then the chain coils and folds into a shape characteristic of the particular protein. The shape of the molecule and the arrangement of atoms or groups on the surface give it the properties specific to the function it carries out in the organism. The interactions which hold the shape in place can be of four types. Ionic bonds form between positive and negative parts of the chain. New covalent links occur in a few places along the structure. Weak hydrogen bonds are important for many proteins. Hydrophobic attractions also play a part in organising the shape. In a relatively small proportion of proteins and other biological molecules, coordinate links between a metal, such as iron, and groups with oxygen or nitrogen atoms are important. This chapter will discuss and compare these various interactions.

■ 3.2 IONIC BONDING

The sharing of electrons between atoms in pairs, or the distribution of electrons to bonding molecular orbitals, gives rise to neutral molecules with covalent bonds linking the atoms together (section 2.5). Strong chemical bonds can be formed in a different way, by the complete transfer of an electron, or electrons, from one atom to another. It is usually a metal that donates the electron and a non-metal that receives it. Often the metal which provides the electrons has only a few electrons in the outer electron shell. It gives all of these to the non-metal which receives enough electrons to fill its already near-complete electron shell. Thus the metal is left with an empty outer shell and a filled inner shell immediately below it. The non-metal atom achieves a full outer shell of electrons.

Since an electron carries an electrical charge, the transfer of electrons between atoms leads to a change in the charge on the atom. The atom that loses an electron now has one more proton than the number of electrons (the two were present in equal numbers in the original atom). The extra proton has a positive charge so the particle formed has an overall positive charge and is called a **cation**.

Transfer of an electron from one atom to another forms a positive cation and a negative anion

The atom gaining an electron also gains the negative charge on the electron; it becomes negatively charged and is called an **anion**.

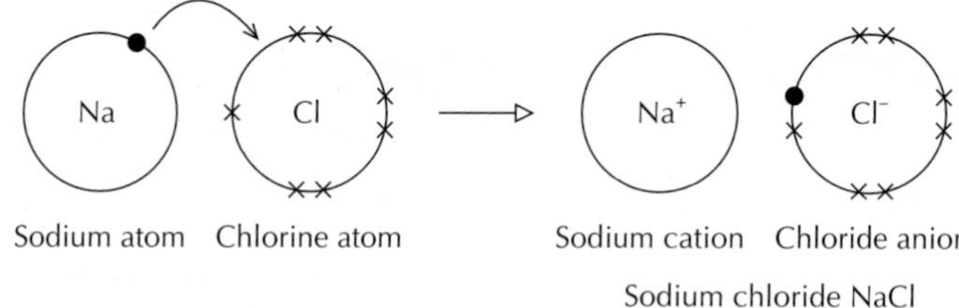

• **Figure 3.1** Electron transfer from sodium to chlorine to form the ionic bond in sodium chloride.

Table 3.1 Comparison of protons, electrons and charges in sodium, chlorine and sodium chloride

	Sodium atom *Na*	*Chlorine atom* *Cl*	*Sodium cation* *Na^+*	*Chloride anion* *Cl^-*
Number of protons	11	17	11	17
Number of electrons	11	17	10	18
Electrical charge	0	0	+1	–1

Sodium chloride, NaCl, is involved in the ion balance across membranes, and is an ionic compound formed by electron transfer. The electron configurations of the two atoms are: Na 2.8.1 Cl 2.8.7

The single electron in the outer shell of sodium is transferred to the near-complete outer shell of chlorine. The electron configurations then become: Na^+ 2.8 Cl^- 2.8.8

Cations and anions usually have full outer electron shells

It is clear that each ion, the sodium cation, Na^+, and the chloride anion, Cl^-, has a set of complete electron shells. The plus sign written after the symbol for sodium and above the line indicates that the cation has a positive charge. In the same way, the minus written after the symbol Cl shows the negative charge on the anion. The + and – charges exactly balance one another, so the compound sodium chloride is neutral overall.

It must be remembered that only an electron is transferred between the atoms; the protons and neutrons within each atom remain fixed. Electron transfer and the formation of ions can be shown using an outer shell electron diagram (Figure 3.1).

Unlike covalent bonding, the outer electron shells of the two ions do not overlap. The opposite charges on the ions attract one another strongly so that the two ions are held close together in the compound. This is an **ionic bond**. The number of protons and electrons in the atoms and ions involved in forming sodium chloride are shown in Table 3.1.

WORKED EXAMPLE 3.1

Calcium chloride, $CaCl_2$, provides a source of soluble calcium for marine organisms. Show the ionic bonding in the compound using electron structure diagrams.

ANSWER

Write the electron configurations of the atoms (Table 2.1). Ca 2.8.8.2 Cl 2.8.7

More than one electron may be transferred in the formation of ions

From the electron structure of the atoms and the formula, $CaCl_2$, calcium will transfer one electron to each of the two chlorine atoms, to leave full outer electron shells for the calcium cation and the two chloride anions. Ca^{2+} 2.8.8 2 × Cl^- 2.8.8

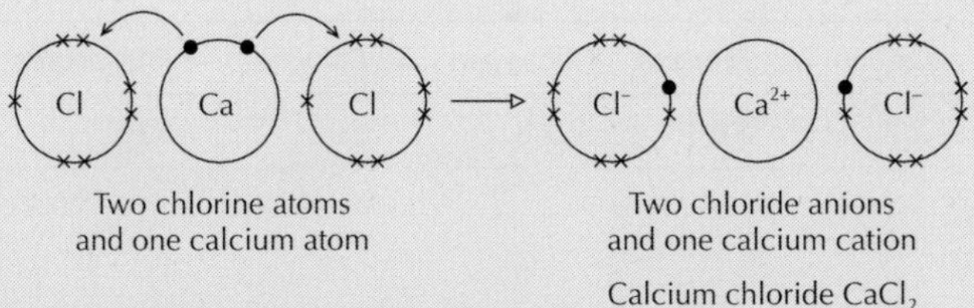

• **Figure 3.2** Electron transfer from calcium to chlorine to form two ionic bonds in calcium chloride.

Draw the outer shell electron diagrams for the atoms and ions (Figure 3.2).
The ionic bonds are formed by the attraction between the calcium ion and the two chloride ions.

QUESTION 3.1

Use electron configurations (Table 2.1) and electron structure diagrams to show the formation of ions and ionic bonding in potassium chloride, KCl, which is implicated in the anion–cation balance across the neuron membrane.

3.3 POLAR COVALENT BONDS

The formation of a chemical bond between atoms has been described in two different ways. In the formation of a covalent bond, a pair of electrons is shared equally between two atoms. The outer electron shells overlap and the pair of electrons is concentrated in the region of space between the two atoms. Alternatively, when an ionic bond is formed, an electron is transferred completely from one atom to another to form a positive cation and a negative anion that are attracted strongly to one another by electrostatic forces.

Electrons may be shared unequally in a covalent bond

These bond types appear to be quite distinct and mutually exclusive. But this is not strictly the case. It is possible in a covalent bond for the electrons to be shared unequally, with one atom achieving more than a half-share of the electron pair. Within an ionic compound, the electrons might not be transferred completely from one atom to another. A study of many biological molecules reveals that most of them fail to show either pure covalent or pure ionic bonding but have intermediate character.

A covalent bond is formed by equal sharing of two electrons only when the two elements bound together are the same. For example, the bond in the hydrogen molecule, H_2, is a pure covalent bond. When the elements are different, the electron pair will be unequally shared. The extent of uneven electron sharing in the bond is determined by the differences in electronegativity between the elements (section 1.6). The element with the higher electronegativity will gain the greater share of the electrons.

Numerical values of electronegativity have been assigned to the elements. Some of those useful to the life sciences are collected in Table 3.2. When we compare the differences between the values for a pair of elements, the relative inequality of electron sharing in a bond can be seen. In a carbon–hydrogen bond, C—H, the electronegativity difference is $2.5 - 2.1 = 0.4$ (Table 3.2). Thus the bond has only a small drift of electrons. These move towards the element with the higher value. So carbon has a very small extra share of the electron pair. The term **polarisation** is used to describe a bond with an unequal sharing of electrons. The carbon–hydrogen bond is only slightly polarised.

The electrons in a polar bond are not shared equally

The oxygen–hydrogen bond, O—H, has an electronegativity difference of $3.5 - 2.1 = 1.4$ (Table 3.2), more than three times that for the carbon–hydrogen bond. Thus the

Table 3.2 Electronegativity values for some elements of biological importance

Hydrogen	*Carbon*	*Nitrogen*	*Oxygen*	*Chlorine*	*Phosphorus*	*Sulphur*
2.1	2.5	3.0	3.5	3.0	2.1	2.5

electrons in the oxygen–hydrogen bond are strongly attracted by the oxygen. The bond shows high polarisation.

It is common to express the polar nature of the bond and the direction of polarisation by a partial electrical charge placed near the symbol for the element in a bond. A partial charge is shown by Greek δ (delta). In this way the oxygen–hydrogen and nitrogen–hydrogen bonds are written:

$$\overset{\delta-}{\mathrm{O}}-\overset{\delta+}{\mathrm{H}} \qquad \overset{\delta-}{\mathrm{N}}-\overset{\delta+}{\mathrm{H}}$$

WORKED EXAMPLE 3.2

Use numerical values to determine the electronegativity difference in the carbon–nitrogen bond, C—N. Comment on the differences in terms of bond polarisation.

ANSWER

Write the electronegativity values (Table 3.2) and subtract them to give the electronegativity difference.

C 2.5 N 3.0 3.0 − 2.5 = 0.5

The electronegativity difference in the carbon–nitrogen bond is small and so the bond is only slightly polarised.

QUESTION 3.2

(a) What are the differences in electronegativity between the elements making up the sulphur–oxygen bond?

(b) Which atom will be positively and which negatively polarised?

Bond polarity has important consequences for the reactions of biologically significant molecules (Chapters 7 and 8) and for the structure of molecules, especially proteins.

■ 3.4 DIPOLE-DIPOLE FORCES

Molecules with dipolar bonds attract one another

The uneven sharing of electron pairs in covalent bonds because of electronegativity differences has been discussed (section 3.3). We have seen that a covalent bond such as the oxygen–hydrogen, O—H, bond in water is permanently polarised with positive and negative ends. The bond is **dipolar**. Molecules with polar bonds attract one another. The positive end of a bond in one molecule attracts the negative part of another bond in a second molecule. Carbon dioxide has polar carbon–oxygen bonds. The molecules attract one another and cause solid carbon dioxide to melt at a considerably higher temperature than expected. Dipolar–dipolar forces are strongest when a small group of highly electronegative elements are bound to hydrogen. This will be considered in section 3.5 on the hydrogen bond.

WORKED EXAMPLE 3.3

Trichloromethane, $CHCl_3$, is a tetrahedral molecule with polar carbon–chlorine bonds. Use diagrams to illustrate the dipole–dipole interactions between molecules.

ANSWER

(a) Draw the structural formula of the molecule.

(b) Add the relative partial charges based on the electronegativity values (Table 3.2).

(c) Arrange two molecules to allow opposite ends of the dipoles to interact. Show the interaction with a broken line (Figure 3.3).

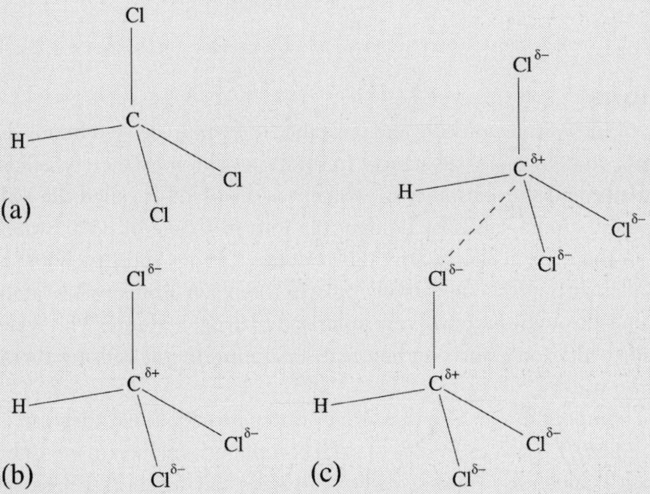

• **Figure 3.3** Dipole–dipole interaction in trichloromethane.

QUESTION 3.3

Draw diagrams to show the dipole–dipole attraction between two molecules of the respiration product carbon dioxide, CO_2.

3.5 THE HYDROGEN BOND

The hydrogen atom is unusual in that it has only a single electron and its nucleus consists of a single proton. When hydrogen forms a bond to a highly electronegative atom such as oxygen (section 3.3), the electrons in the bond are attracted to the oxygen atom. This leaves the hydrogen nucleus (a proton) with only a thin shield of electron density around it, giving the positive end of a dipole. The positive charge carried on the nucleus is attracted strongly by the negative charge of lone pair electrons (section 2.9) on the oxygen atom of a neighbouring molecule (Figure 3.4). These electrons form the negative end of a dipole. A common compound which contains oxygen–hydrogen bonds and shows this effect is water, H_2O. The attractive force is called the **hydrogen bond**. Water molecules in the liquid phase are linked by hydrogen bonding in a three-dimensional array because each has two hydrogen atoms and two lone pairs on oxygen.

• **Figure 3.4** Hydrogen bond formed between oxygen and hydrogen covalently linked to oxygen.

$$\overset{\delta-}{O}-\overset{\delta+}{H}\cdots\overset{\delta-}{O}$$

Other electronegative elements, in particular nitrogen, form hydrogen bonds to compounds containing N—H groups and to those with O—H groups:

$$\overset{\delta-}{\mathrm{N}}—\overset{\delta+}{\mathrm{H}}\cdots\overset{\delta-}{\mathrm{N}} \qquad \overset{\delta-}{\mathrm{N}}—\overset{\delta+}{\mathrm{H}}\cdots\overset{\delta-}{\mathrm{O}} \qquad \overset{\delta-}{\mathrm{O}}—\overset{\delta+}{\mathrm{H}}\cdots\overset{\delta-}{\mathrm{N}}$$

Hydrogen bonding is of considerable importance for amines.

The existence of the hydrogen bond has a number of significant consequences. The low relative molecular mass of water (18) suggests that it should be a gas at room temperature. We know, of course, that it is a liquid; hydrogen bonding makes it much less volatile than expected. The α-helix and β-pleated sheet structures of proteins depend on extensive hydrogen bonding. Base pairing in the DNA helix requires hydrogen bonding.

Hydrogen bonding stabilises protein structures

■ SOLUTIONS

The polarity of the water molecule and its ability to form hydrogen bonds determine the ability of water to hold many substances in solution. The substance which is dissolved is called the **solute** and the substance in which it is dissolved is called the **solvent**.

Water dissolves ionic solids by holding the ions in a cage of polar water molecules. The picture in the margin shows how this is done. Cations (positive ions) are bound to water molecules through the negatively polarised oxygen atoms, while anions (negative ions) are bound through the positively polarised hydrogen atoms. The water molecules can thus bind to the ions, surrounding and separating them, causing them to disperse throughout the water.

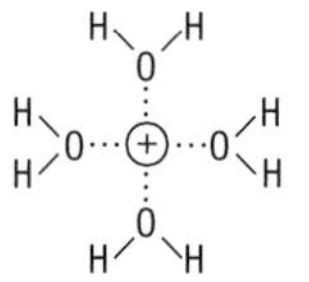

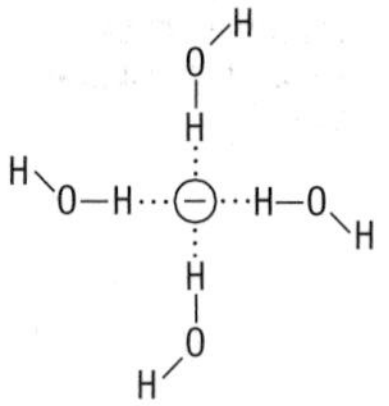

Water molecules can solvate both cations and anions

Non-ionic compounds can also dissolve in water provided that they can participate in hydrogen bonding. Typical examples of such compounds include many of biological significance, such as alcohols, amines, carboxylic acids and sugars (see Chapters 6 and 7). These compounds contain O—H or N—H bonds which are similar to the O—H bonds in water itself. An example of an alcohol hydrogen bonding with water molecules is shown in the margin.

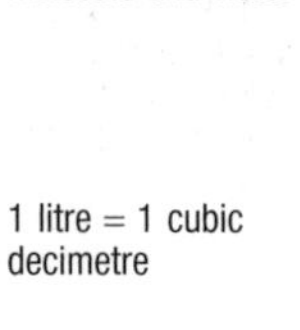

Hydrogen bonding between an alcohol molecule and water

Whether the water molecules are interacting with an ion or with a polar compound, they are still able to hydrogen bond with further water molecules through their unbound ends. The result is that substances dissolved in water are surrounded by a shell of water molecules bound directly to them or to other water molecules which are directly bound to them. This shell of water molecules is called the **hydration sphere** of the substance. Individual water molecules enter and leave the hydration sphere continuously, but the ion or molecule remains surrounded at all times.

The concentration of a solution can be measured in various units, such as grams per litre or kilograms per cubic metre. However, the most common and useful measure for chemical purposes is moles per litre, or moles per cubic decimetre. A litre and a cubic decimetre are the same thing.

1 litre = 1 cubic decimetre

QUESTION 3.4

The amino acid glycine, $H_2N—CH_2—COOH$, shows hydrogen bonding between molecules. Draw structures to show (a) N · · · H—O and (b) O—H · · · O hydrogen bonding between two molecules. Indicate the lone pair electrons involved in each case.

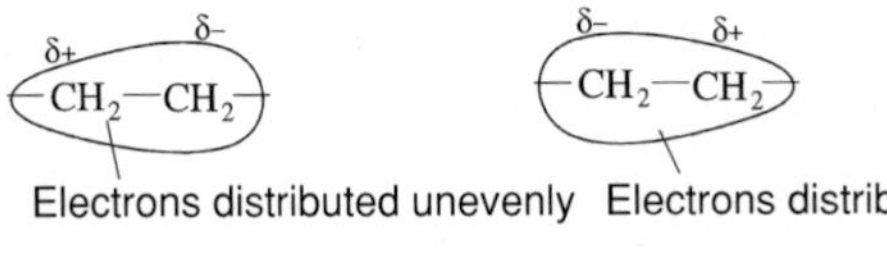

• **Figure 3.5** Temporary polarisation of electrons in an alkyl chain. (a), (b) and (c) show possible electron distributions at different moments of time.

■ 3.6 VAN DER WAALS FORCES

Molecular elements such as oxygen have no dipolar character in the oxygen–oxygen bond. But these molecules have a weak attraction for one another. This arises because the electrons within the molecule are in rapid motion. At any instant of time the electrons can be unevenly distributed, causing one end of the molecule to be very slightly positive and the other end slightly negative. This is shown in Figure 3.5 for a carbon–carbon bond in an alkyl chain. This instantaneous dipole can then attract a similar dipole in a second molecule. When considering these very weak forces, it is necessary always to bear in mind that they are temporary, not permanent, and that they operate over quite short distances. It is easier for a large number of electrons in a big molecule to be unevenly spread than it is for a few electrons in a small molecule to be perturbed. Thus van der Waals forces are more significant in large molecules than in small ones. Within a folded protein molecule, uncharged non-polar groups are packed tightly together. van der Waals attractions between them contribute to the stability of the structure. Lipid molecules contain long non-polar chains. These mutually attract one another when the chains are closely aligned and optimally spaced. This helps to stabilise the bilayer or micelle structure adopted by these substances (see Figure 3.6).

Weak van der Waals attractions stabilise close-packed non-polar groups

■ 3.7 THE HYDROPHOBIC EFFECT

A large biological molecule like the protein haemoglobin has several non-polar regions within it. When the molecule is unfolded and dissolved in water the non-polar regions are in close contact with water. These groups repel water molecules which organise themselves to form a cage-like structure around each region. This ordered arrangement is unfavourable; it has a low level of randomness or entropy (sections 11.11 and 12.11). Folding of the molecule takes place so that the non-polar regions are buried within the structure leaving only polar groups in contact with water. The organised groups of water molecules which surrounded the non-polar regions are released and become less organised. The level of randomness of the whole protein–water solution increases, as does the entropy. An increase in the randomness of a structure increases its stability. This is the **hydrophobic effect**; it is not a bonding interaction but provides a significant factor in the overall stabilisation of a globular protein. The bilayer structure of lipid membranes (Chapter 7) is stabilised in the same way by the hydrophobic effect. The relative strengths of covalent and non-covalent bonds are shown in Figure 3.7. The non-covalent bond energies are about an order of magnitude weaker than those in covalent bonds.

Non-polar groups in large molecules are stabilised by entropy factors – the hydrophobic effect

(a)
δ+ CH_2 δ− CH_2
CH_2 CH_2
δ− CH_2 δ+ CH_2

(b)
CH_2 CH_2
δ+ CH_2 δ− CH_2
CH_2 CH_2

(c)
δ− CH_2 δ+ CH_2
δ+ CH_2 δ− CH_2
CH_2 CH_2

• **Figure 3.6** Possible temporary polarisation of $-CH_2-$ groups in two adjacent alkyl chains within lipid molecules. (a), (b) and (c) show polarisation at different moments of time.

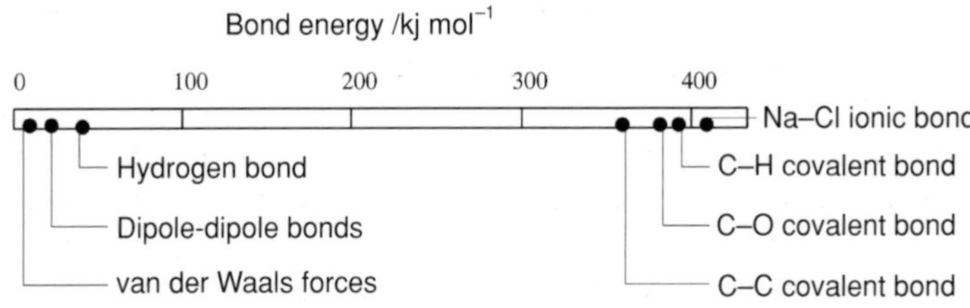

• **Figure 3.7** Comparison of covalent, ionic and non-covalent bond energies.

■ 3.8 COORDINATE BONDS

Metals can form covalent bonds by a different process to non-metals. Thus when two non-metal hydrogen atoms form a covalent bond each atom contributes one electron. Metals, and especially transition metal ions, take part in covalent links by receiving two electrons from a suitable non-metal donor. The electrons provided are usually from a lone pair on an element such as nitrogen, oxygen or sulphur. A compound that provides a pair of electrons in this way is called a **ligand**. In biological ligands the donor atom may be nitrogen in a five-membered ring structure such as histidine or the oxygen of a carboxylate group in glutamic acid. Transition metal ions can receive pairs of electrons because they have available vacant orbitals within the valence shell. Usually four or six pairs of electrons are donated by the ligands. These extra electrons will often complete the filling of the valence shell orbitals and give the metal ion a stable, full outer shell, electron configuration. The covalent bond formed in this way is called a **coordinate bond**.

A metal receives pairs of electrons into vacant orbitals from non-metals to form coordinate bonds

Although only a few of the huge range of biological compounds contain a metal, those that do are often of vital importance to the organism. Haemoglobin and the closely related protein myoglobin illustrate the role and significance of the metal ion in the structure. Within higher animals oxygen must be transported for cell respiration. The iron atoms in haemoglobin perform this function by binding free oxygen, transporting it unchanged to the required site and giving it up on demand.

The structure holds an iron(II), Fe^{2+}, ion within a porphyrin ring which has four nitrogen ligands in a rigid plane around the ion. Below the plane is a histidine group with a fifth nitrogen ligand, while the oxygen molecule occupies the sixth coordination position and behaves like a ligand (Figure 3.8). Within the protected, hydrophobic environment of the haemoglobin molecule, the iron(II) is able to bind and hold oxygen without itself becoming oxidised as might be expected.

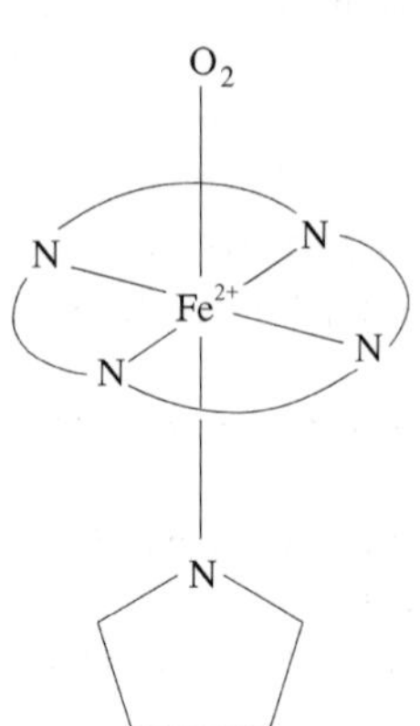

• **Figure 3.8** Oxygen coordinated to iron(II) in the octahedral environment provided by the haemoglobin molecule.

ADVANCED TOPIC BOX: CHELATES

The majority of ligands, such as ammonia, donate one electron pair to a metal and are known as monodentate (one-toothed). Some ligands contain two or more atoms that are capable of forming co-ordinate bonds with the same metal ion. Such ligands are called **chelate** ligands. The simplest chelates have two atoms which act as ligands and are called bidentate (two-toothed). Examples include ethylene-1,2-diamine with the two amines capable of complex formation, and the carboxylate anion where the two oxygen atoms can chelate a metal ion (see margin). Many multidentate ligands exist such as the tetradentate porphyrin ring of the haem group shown in figure 3.8. The compound ethylenediamine tetraacetic acid (EDTA), shown in the margin, is a widely used hexadentate chelator where one molecule of EDTA is capable of co-ordinating a metal atom octahedrally. Chelators are very effective at co-ordinating metal ions, probably as at least two monodentate ligands are freed into solution as the chelate binds. This increases the entropy within the system.

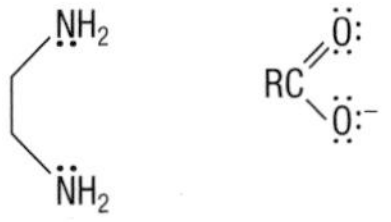

NH_2 … NH_2 ; $RC(=\ddot{O}:)\ddot{O}:^-$

Some bidentate ligands

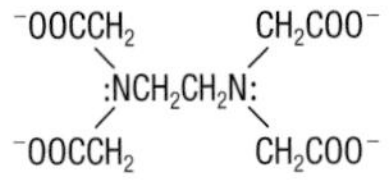

$^-OOCCH_2$ CH_2COO^-
$:NCH_2CH_2N:$
$^-OOCCH_2$ CH_2COO^-

the hexadentate chelator EDTA

■ SUMMARY

An ionic bond is formed when one or more electrons are transferred completely from a metal to a non-metal. The metal forms a positive cation and the non-metal forms a negative anion. These two oppositely charged particles are held together strongly by electrostatic attraction to give an ionic bond. Both ions achieve a full outer shell of electrons and are stable. Pure covalent and ionic bonds are exceptional. It is usual for a covalent bond to have the electron pair which comprises the bond unequally distributed between the two atoms involved. An ionic bond often has less than complete electron transfer. The bonds that result are polar, with a charge separation across the bond. Electronegativity differences determine the direction of bond polarisation. Hydrogen atoms bound to strongly electronegative elements have unusual properties. They are slightly positive and are attracted to electronegative atoms in other molecules to form weak hydrogen bonds. Other molecules with polar bonds are attracted to one another by quite weak electrostatic forces; this is dipole–dipole interaction. Non-polar groups in molecules are very weakly attracted to one another by van der Waals forces. Hydrophobic groups cluster tightly together in large biological molecules by the hydrophobic effect. This relies on changes in randomness on folding or organising molecules in solution in water. Transition metal ions are present in a range of biologically active molecules. They are held in place by several coordinate bonds and play a specific role in metabolism.

■ END OF CHAPTER QUESTIONS

Question 3.5 Magnesium chloride, $MgCl_2$, provides a source of soluble magnesium for organisms.

(a) Use outer electron shell diagrams to show how ionic bonds are formed in this compound.

(b) List the number of protons and electrons in the magnesium and chlorine atoms and ions.

Question 3.6 For each of the polar bonds phosphorus–oxygen, carbon–oxygen and carbon–chlorine:

(a) Find the electronegativity difference (Table 3.2).

(b) Which bond is the most polar?

(c) Write partial positive and negative charges over the bond formulae to indicate the direction of polarisation.

Question 3.7 Several types of hydrogen bonding are possible in the amino acid serine:

$$\begin{array}{c} H_2N\underset{|}{C}HCOOH \\ CH_2OH \end{array}$$

Use structural formulae to show hydrogen bonding between two molecules of serine for the types listed:

(a) C=O · · · H—O (b) O—H · · · O—H

(c) C=O · · · H—N (d) N—H · · · N

Question 3.8 Give examples of two biological macromolecules which show hydrogen bonding and explain how the bonding is important to the structure of the molecule.

Question 3.9 Explain the statement: '*The hydrophobic effect causes non-bonding regions of protein molecules to cluster together and stabilises the structure even though it is not a bonding effect.*'

Question 3.10 (a) What types of elements participate in forming coordinate bonds?

(b) How is a coordinate bond different from a normal covalent bond?

(c) Describe the function and significance of the coordinate bonds in the haemoglobin molecule.

CHEMICAL REACTIONS

■ 4.1 INTRODUCTION

Biochemical and chemical reactions are vital to living organisms. For example, complex molecules such as starch or proteins have to be broken down chemically into simpler molecules. These simpler molecules can be consumed in further biochemical reactions to produce energy or can be used as building blocks to construct the complex molecules that organisms need for growth, repair and reproduction. In fact, all the processes of metabolism involve biochemical and chemical reactions and so knowledge of such reactions is necessary for the understanding of living organisms.

■ 4.2 RATES OF REACTION

The speed at which biochemical reactions occur can vary enormously. Some chemical reactions, such as explosions, take place in thousandths or even millionths of a second, whereas others, such as the rusting of iron, can take years or decades to complete. For living organisms the rate of reaction is extremely important. The smooth running of an organism's metabolism requires that reactions should always go at an optimum rate.

Chemical reactions can vary enormously in rate

■ 4.3 FACTORS AFFECTING RATE OF REACTION

Reactions between molecules can only occur when the molecules collide. If the collision is sufficiently energetic, bonds can be broken and then reformed in a different arrangement. The molecules which exist originally and take part in the reaction are termed the reactants. The molecules that result after reaction has taken place are termed the products.

Chemical reactions can occur when molecules collide

Three factors are important in controlling the rate of a reaction:

1 **Temperature**. Temperature affects the speed at which molecules move: the higher the temperature, the greater is the average speed of the molecules. Therefore, as the temperature increases, collisions between molecules become more violent. This increases the chances that bonds between atoms will be broken during a collision and so increases the chances of a reaction occurring. This topic is dealt with in greater depth in Chapter 12.

Increasing the temperature will always increase the rate of a reaction

2 **Catalysis**. A catalyst is a substance which increases (or sometimes decreases) the rate of a reaction while not itself being chemically changed at the end of the reaction. In biological systems, catalysis is brought about by **enzymes**. All reactions in living things are controlled by enzymes. This topic is covered in greater depth in Chapter 12.

Enzymes control the rates of reactions in living organisms

Increasing the concentration of reactants often increases the rate of a reaction

3 **Concentration**. The higher the concentration of a reactant in a mixture, the greater is the number of collisions that molecules of that substance will undergo in a given time. The greater number of collisions leads to a faster rate of reaction. In this chapter we examine the role of concentration in determining the rate of reaction.

■ 4.4 RATE EQUATIONS

Rate equations show how the rate of a reaction depends on the concentration of the reactants

For a general reaction such as:

$$A + B \rightarrow C + D$$

A and B are termed the **reactants** while C and D are called the **products**. It is commonly found that at a fixed temperature and with a constant amount of catalyst present, the rate of the reaction (i.e. the speed at which reactants are turned into products) can be modelled by an equation such as:

$$\text{Rate} = -k[A]^x[B]^y \qquad (4.1)$$

where k is a constant called the **rate constant**, [A] and [B] are the concentrations of A and B in mol dm^{-3} and the powers to which the concentrations are raised, x and y, are chosen to match the observed rate of reaction. The powers x and y are called the **orders of reaction** with respect to reactants A and B, respectively. They are usually small whole numbers such as 0, 1 or 2. The **overall order of reaction** is given by $x + y$. The minus sign in Equation 4.1 indicates that as time goes on the concentration of reactants decreases.

■ 4.5 INTEGRATED FORMS OF RATE EQUATIONS

Integrated rate equations allow us to determine rate constants

The rate equation above (called the **differential form** of the rate equation) allows us to calculate the rate of a reaction at any time provided we know the concentrations of the reactants at that time and the rate constant. Often we can measure the concentration of a reactant fairly easily, but do not know the rate constant and so cannot use this equation.

Using calculus, it is possible to integrate the rate equation to obtain an equation from which we can calculate the rate constant. We can then use this knowledge to calculate either the rate of reaction expected with a given concentration of reactants or the concentration of reactants left after the reaction has proceeded for a given time.

The result of the integration depends on the order of the reaction. The derivations of the integrated rate equations for zero-, first- and second-order reactions are to be found in the Appendix. In the following sections the integrated rate equations are presented and graphical methods of determining rate constants explained.

■ 4.6 ZERO-ORDER REACTIONS

The rate of a zero-order reaction is independent of the reactant concentrations

In a zero-order reaction the power to which the concentrations of the reactants are raised is zero for all reactants. Any number raised to the power zero is equal to 1, so in this case the rate of the reaction does not depend at all on the concentration of the reactants. The rate equation for such a reaction is:

$$\text{Rate} = -k \qquad (4.2)$$

Zero-order reactions are commonly found in cases where catalysis is occurring. If there are lots of reactant molecules present, the catalyst is working as hard as it possibly can – as soon as one reactant molecule has interacted with the catalyst, another one takes its place. Increasing the concentration of the reactant cannot increase the rate of the reaction because the catalyst cannot work any faster. The catalyst is said to be **saturated**.

■ 4.7 INTEGRATED FORM OF THE ZERO-ORDER RATE EQUATION

The zero-order rate equation can be integrated to give:

$$[A]_t = [A]_0 - kt \tag{4.3}$$

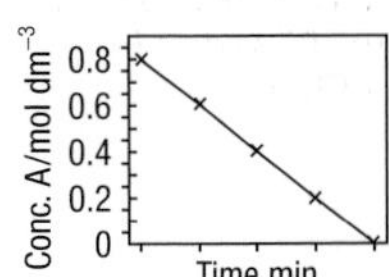

Graph of [A] versus time for a zero-order reaction

where $[A]_0$ and $[A]_t$ are the concentrations of A at the start of the reaction and after a time t has elapsed, respectively.

The integrated form of the rate equation is usually used to determine the rate constant of a reaction. If we can monitor the concentration of A by some means (for example by taking samples for analysis at various intervals) then we can plot a graph of $[A]_t$ versus t. This graph will give a straight line of gradient $-k$ and intercept $[A]_0$.

■ 4.8 FIRST-ORDER REACTIONS

The rate of a first-order reaction is affected by the concentration of only one reactant

In a first-order reaction, the rate at which the reaction goes depends on the concentration of only a single reactant. There may be more than one reactant taking part in the reaction, but only one reactant influences the rate of the reaction. The rate equation for such a reaction is:

$$\text{Rate} = -k[A] \tag{4.4}$$

The reaction is first order with respect to A, while the order with respect to all other reactants is zero. The reaction is thus also first order overall.

As we vary the concentration of A, the rate of reaction will change. For example, if we double the concentration of A from $[A]_1$ to $[A]_2$, so that:

$$[A]_2 = 2[A]_1$$

the rate of reaction becomes:

$$\begin{aligned}\text{Rate} &= k[A]_2 \\ &= 2k[A]_1\end{aligned}$$

The rate of reaction therefore doubles when the concentration of A doubles.

■ 4.9 INTEGRATED FORM OF THE FIRST-ORDER RATE EQUATION

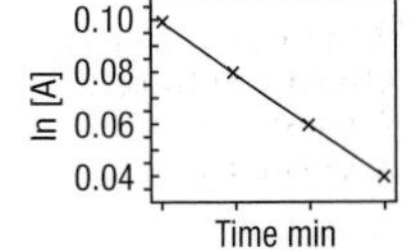

Graph of ln [A] versus time for a first-order reaction

As the reaction proceeds A is used up and so its concentration falls. The rate of the reaction therefore slows as the reaction goes on. The concentration of A at any time during the reaction can be represented by using the integrated form of the rate equation:

$$\ln [A]_t = \ln [A]_0 - kt \tag{4.5}$$

where ln represents natural logarithm, $[A]_0$ and $[A]_t$ are the concentrations of A at the start of the reaction and after a time t has elapsed, respectively.

In this case, we plot a graph of ln $[A]_t$ versus t to obtain a straight line of gradient $-k$ and intercept ln $[A]_0$.

■ 4.10 SECOND-ORDER REACTIONS

Second-order reactions can be of two types:

Type 1. The rate of the reaction depends on the concentration of only one reactant. The rate equation is:

$$\text{Rate} = -k[A]^2 \tag{4.6}$$

As we vary the concentration of A, the rate of reaction will change but in a different way from that seen with first-order reactions. For example, if we double the concentration of A from $[A]_1$ to $[A]_2$, so that:

$$[A]_2 = 2[A]_1$$

then the new rate of reaction will be given by:

$$\begin{aligned}\text{Rate} &= -k[A]_2^2\\ &= k(2[A]_1)^2\\ &= -4k[A]_1\end{aligned}$$

The rate of the reaction therefore increases fourfold when the concentration of A is doubled.

The rate of a second-order reaction can depend on the concentration of either one or two reactants

Type 2. The rate of reaction depends on the concentrations of two reactants. The rate equation is:

$$\text{Rate} = -k[A][B] \tag{4.7}$$

In this case, doubling the concentration of either A or B will double the rate of reaction: doubling both concentrations will increase the rate four-fold.

■ 4.11 INTEGRATED FORMS OF SECOND-ORDER RATE EQUATIONS

Type 1. This rate equation can be integrated to give:

$$\frac{1}{[A]_t} = \frac{1}{[A]_0} + kt \tag{4.8}$$

where $[A]_0$ and $[A]_t$ are the concentrations of A at the start of the reaction and after a time t has elapsed, respectively.

Plotting a graph of $1/[A]_t$ versus t will give a straight line of gradient k and intercept $1/[A]_0$.

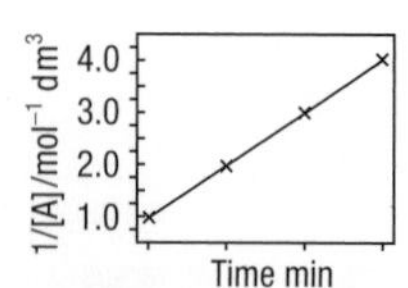

Graph of 1/[A] versus time for a second-order reaction

Type 2. This rate equation can be integrated to give:

$$\frac{1}{[A]_0 - [B]_0} \ln \frac{[B]_0[A]_t}{[A]_0[B]_t} = kt \tag{4.9}$$

where $[A]_0$ and $[B]_0$ are the concentrations of A and B at the start of the reaction and $[A]_t$ and $[B]_t$ are the concentrations of A and B after a time t has elapsed, respectively. To obtain a straight line graph of gradient k from this equation, we would have to plot a graph of the left-hand side of the equation versus t. This would require us to know both $[A]_t$ and $[B]_t$. It is likely to be difficult to obtain simultaneous measurements of the concentrations of two reactants, so a different approach is employed.

Plotting the different graphs to find the one which gives a straight line enables us to determine the order of a reaction

■ 4.12 PSEUDO-FIRST-ORDER REACTIONS

Any reaction the rate of which depends on more than one reactant can be transformed into a pseudo-first-order reaction. In the case of Type 2 second-order reactions, this can be done by having one of the reactants present in large excess. Suppose that we make the concentration of reactant B much higher than the concentration of reactant A. Then, when all of A has reacted, the concentration of B will hardly have changed at all.

Second-order reactions can be made into pseudo-first-order reactions by having one reactant present in high concentration

We have, for the rate equation:

$$\text{Rate} = -k[A][B]$$

If B is present in such a large excess that its concentration hardly changes during the course of the reaction, then [B] is essentially constant, i.e. $[B]_0 = [B]_t$. The rate equation can then be written:

$$\text{Rate} = -C[A]$$

where the new constant $C = k[B]$. The equation now has the same form as that of a first-order reaction, and it can be integrated similarly to give:

$$\ln [A]_t = \ln [A]_0 - Ct \qquad (4.10)$$

A graph of $\ln [A]_t$ versus t will give a straight line of gradient $-Ct = -k[B]_0$, from which k can be calculated if we know the original concentration of B.

WORKED EXAMPLE 4.1

In a reaction between reactants A and B carried out at fixed temperature and constant enzyme concentration the following results were obtained:

[A]/mmol dm^{-3}	[B]/mmol dm^{-3}	Rate of reaction/mmol dm^{-3} min^{-1}
0.2	0.2	1.0×10^{-3}
0.4	0.2	2.0×10^{-3}
0.4	0.4	8.0×10^{-3}

(a) What is the order with respect to each reactant?

(b) What is the overall order?

ANSWER

(a) Doubling the concentration of A while holding the concentration of B constant doubles the rate of reaction, so that the rate is directly proportional to the concentration of A, that is the rate varies according to $[A]^1$. The reaction is therefore first order with respect to A. Doubling the concentration of B while holding the concentration of A constant leads to a fourfold increase in the rate of reaction, so that the rate is proportional to the concentration of B squared, that is the rate varies according to $[B]^2$. The reaction is therefore second order with respect to B.

(b) Since 1 + 2 = 3, the reaction is third order overall.

QUESTION 4.1

Coenzyme A (CoA) can be prepared by the reaction of its thiol derivative, CoASH, with ethanoyl chloride. Under certain conditions the following results were obtained:

[CoASH] /mmol dm^{-3}	[Ethanoyl chloride] /mmol dm^{-3}	Rate of reaction /mmol dm^{-3} min^{-1}
0.01	0.1	3.4×10^{-3}
0.03	0.1	1.02×10^{-2}
0.01	0.2	6.8×10^{-3}

(a) What is the order of reaction with respect to each reactant?

(b) What is the overall order?

■ 4.13 REVERSIBLE REACTIONS

The reactions we have considered so far have been unidirectional, that is reactants have been transformed into products in a one-way process. However, in many cases it is found that the products of a reaction can also react together to regenerate reactants. Such a reaction is said to be **reversible** and is written with a double-headed arrow:

$$A + B \rightleftharpoons C + D$$

■ 4.14 EQUILIBRIUM

If we start off such a reaction by mixing reactant A with reactant B they start to react to form C and D. At first, only the forward reaction can occur, since no C or D has yet been produced, but as C and D begin to accumulate, so they begin to react together to produce A and B. As A and B react, their concentration decreases and so the rate of the forward reaction slows down. Conversely, as C and D are produced, their concentration increases and the rate of the backward reaction rises. This continues until the rates of the forward and backward reactions become equal. At this stage the reaction is said to be in **equilibrium**. A and B continue to react together to form C and D, but C and D are now reacting at an equal rate to produce A and B.

The rate of the forward reaction is given by a rate equation of the form:

$$\text{Rate(f)} = k_f[A][B]$$

where Rate(f) is the rate of the forward reaction, k_f is the rate constant of the forward reaction and [A] and [B] are the concentrations of A and B, respectively.

The rate of the backward reaction is given by a similar rate equation:

$$\text{Rate(b)} = k_b[C][D]$$

where Rate(b) is the rate of the backward reaction, k_b is the rate constant for the backward reaction and [C] and [D] are the concentrations of C and D, respectively.

When the reaction has reached equilibrium, the forward and backward rates are equal:

$$k_f[A][B] = k_b[C][D]$$

Rearranging this, we obtain:

$$\frac{k_f}{k_b} = \frac{[C][D]}{[A][B]} \qquad (4.11)$$

The left-hand side of this equation is formed from two constants and therefore must itself be constant. Since this constant is equal to the right-hand side, the right-hand side must also be a constant. This constant is called the **equilibrium constant**, K_{eq}, for this reaction.

$$K_{eq} = \frac{k_f}{k_b} = \frac{[C][D]}{[A][B]} \qquad (4.12)$$

Note that upper case letters are used for equilibrium constants, while lower case letters are used for rate constants. The higher the equilibrium constant, the further the reaction goes in the forward direction.

WORKED EXAMPLE 4.2

ADP reacts with inorganic phosphate (P_i) to produce ATP according to the equation:

$$\text{ADP} + P_i \rightleftharpoons \text{ATP}$$

Under certain conditions the concentrations of ADP, P_i and ATP in intracellular fluid were found to be 4.2×10^{-3} mol dm^{-3}, 6.8×10^{-4} mol dm^{-3} and 1.85×10^{-11} mol dm^{-3}, respectively. (a) Write the expression for the equilibrium constant, K_{eq}, for this reaction. (b) Calculate the equilibrium constant for the reaction.

ANSWER

(a) The equilibrium constant, K_{eq}, is given by the following equation, where the concentrations of the products are divided by the concentrations of the reactants:

$$K_{eq} = \frac{[\text{ATP}]}{[\text{ADP}][P_i]}$$

(b) Introduce the numerical values for the concentrations of the reactants and products into the equation:

$$K_{eq} = \frac{1.85 \times 10^{-11}\ \text{mol dm}^{-3}}{4.2 \times 10^{-3}\ \text{mol dm}^{-3} \times 6.8 \times 10^{-4}\ \text{mol dm}^{-3}}$$

$$= 6.47 \times 10^{-6}\ \text{mol}^{-1}\ \text{dm}^{3}$$

QUESTION 4.2

Glucose-1-phosphate (G1P) is converted into glucose-6-phosphate (G6P) by the enzyme phosphoglucomutase according to the equation:

$$\text{G1P} \rightleftharpoons \text{G6P}$$

At equilibrium it is found that the concentration of G1P = 3.0×10^{-3} mol dm^{-3} and the concentration of G6P = 5.8×10^{-2} mol dm^{-3}. (a) Write the expression for the equilibrium constant for this reaction. (b) Calculate the equilibrium constant for the reaction.

WORKED EXAMPLE 4.3

In the self-dissociation reaction of water,

$$H_2O \rightleftharpoons H^+ + OH^-$$

the forward reaction has a rate constant = 2.43×10^{-5} s^{-1} and a backward rate constant = 1.35×10^{11} mol^{-1} dm^3 s^{-1}. (a) Write the equation for the equilibrium constant K_{eq}. (b) Calculate the equilibrium constant for this reaction.

ANSWER

(a) $K_{eq} = \dfrac{k_f}{k_b}$

(b) Introduce the values for the rate constants into the equation:

$$K_{eq} = \frac{2.43 \times 10^{-5}\ \text{s}^{-1}}{1.35 \times 10^{11}\ \text{mol}^{-1}\ \text{dm}^3\ \text{s}^{-1}}$$

$$= 1.8 \times 10^{-16}\ \text{mol dm}^{-3}$$

QUESTION 4.3

The coenzyme NADH associates with the enzyme alcohol dehydrogenase (ADH) according to the equation:

$$\text{NADH} + \text{ADH} \rightleftharpoons \text{NADH–ADH}$$

The forward rate constant for this reaction is = 1.3×10^7 mol^{-1} dm^3 s^{-1} and the backward rate constant = 3.2 s^{-1}. (a) Write the equation for the equilibrium constant, K_{eq}. (b) Calculate the equilibrium constant for the association reaction.

■ SUMMARY

The rate of a biochemical reaction often depends on the concentration of one or more reactants. The differential rate equation expresses this dependence in mathematical form. The integrated form of the rate equation allows us to determine rate constants and orders of reaction.

Some reactions are reversible, giving rise to reactions which do not go to completion, but instead approach equilibrium, in which forward and backward rates of reaction are equal. Such equilibria are characterised by an equilibrium constant which contains information about how far the reaction proceeds.

■ SUGGESTED FURTHER READING

Price, N.C. and R.A. Dwek (1989) *Principles and Problems in Physical Chemistry for Biochemists*, Ch. 9, 2nd edn. Oxford University Press.

■ END OF CHAPTER QUESTIONS

Question 4.4 In a reaction involving two reactants, A and B, the following results were obtained:

[A] /mmol dm^{-3}	[B] /mmol dm^{-3}	Rate /mmol dm^{-3} min^{-1}
0.01	0.1	2.86×10^{-5}
0.01	0.2	1.14×10^{-4}
0.02	0.1	2.86×10^{-5}

(a) What is the order with respect to each reactant?

(b) What is the overall order?

Question 4.5 Plants convert glucose to ascorbic acid (vitamin C). One step in the mechanism involves the ionisation of gluconic acid:

$$C_5H_{11}O_5COOH \rightleftharpoons C_5H_{11}COO^- + H^+$$

$$\text{gluconic acid} \rightleftharpoons \text{gluconate anion} + \text{hydrogen ion}$$

At equilibrium it is found that the concentrations present are:

$$[C_5H_{11}O_5COOH] = 3.2 \times 10^{-2}\ \text{mol dm}^{-3}$$

$$[C_5H_{11}O_5COO^-] = 6.6 \times 10^{-4}\ \text{mol dm}^{-3}$$

$$[H^+] = 6.6 \times 10^{-4}\ \text{mol dm}^{-3}$$

(a) Write down the expression for the equilibrium constant, K_{eq}, for this reaction.

(b) Calculate the equilibrium constant.

Question 4.6 In an equilibrium reaction:

$$A \rightleftharpoons B$$

The rate constant of the forward reaction is $7.2 \times 10^{-3}\ s^{-1}$ while the rate constant of the backward reaction is $2.2 \times 10^{-5}\ s^{-1}$. Calculate the equilibrium constant for this reaction.

5 ACIDS, BASES AND BUFFERS

■ 5.1 INTRODUCTION

Biochemical reactions occur in aqueous solution. Many biochemicals are sensitive to the level of acidity or basicity of the water. Enzyme reactions in particular are susceptible to changes in acidity, and so many organisms seek to maintain internal levels of acidity at optimum values. In this chapter we examine what is meant by terms such as acid and base, how the degree of acidity is measured, and how we can use buffers to maintain a desirable level of acidity or basicity.

■ 5.2 IONISATION OF WATER

Water is not always in the form H_2O as shown in Figure 2.3. About one molecule in five hundred million in pure water is ionised:

$$H_2O \rightleftharpoons H^+ + OH^-$$

$$\text{water} \rightleftharpoons \text{proton} + \text{hydroxide ion}$$

Small though the extent of this ionisation is, it has profound effects on the properties of water.

The equilibrium constant for the above reaction (section 4.13, worked example 4.3) is given by:

$$K_{eq} = \frac{[H^+][OH^-]}{[H_2O]}$$

The concentration of water ($[H_2O]$) in pure water is about 55 mol dm^{-3}. Changes in the concentration of water due to changes in the proton concentration ($[H^+]$) are so small compared to the total concentration of water that the term $[H_2O]$ can be considered to be constant. Instead of using the equilibrium constant, it is usual to use the ionisation product, K_w:

$K_w = [H^+][OH^-] = 10^{-14}$

$$K_w = [H^+][OH^-]$$

In pure water, $K_w = 10^{-14}$.

■ 5.3 THE HYDROGEN ION

The hydrogen ion is often written H^+. This implies that the hydrogen ion is a naked proton. Such a species does not exist under normal circumstances (naked protons can only exist in high vacuum) because its enormous charge density would rapidly strip electrons from any surrounding matter.

The hydrogen ion which exists in solution is best represented by the hydronium ion, H_3O^+, and the autoionisation of water represented by the equation:

$$2H_2O \rightleftharpoons H_3O^+ + OH^- \quad (5.1)$$

$$\text{water} \rightleftharpoons \text{hydronium ion} + \text{hydroxide ion}$$

The hydronium ion would in reality be further solvated.

Despite these facts, it is very common to use H^+ in equations, and to refer to acids as 'proton donors' (see below).

■ 5.4 ACIDS AND BASES

There are several definitions in use as to what exactly acids and bases are. For our purposes the most useful definitions are those due to Brønsted and Lowry:

An acid is a proton (H^+) donor
A base is a proton acceptor

Thus, an acid is a substance which produces hydrogen ions in a solution, while a base is a substance which removes hydrogen ions from a solution.

■ 5.5 STRONG ACIDS AND STRONG BASES

Strong acids and bases are completely ionised in solution. Strong acids dissolve to produce hydrogen ions in solution:

$$HCl + H_2O \rightarrow H_3O^+ + Cl^-$$

$$\text{hydrogen chloride} + \text{water} \rightarrow \text{hydronium ion} + \text{chloride ion}$$

Other commonly encountered strong acids include nitric acid, HNO_3, and sulphuric acid, H_2SO_4.

Strong bases dissolve to produce hydroxide ions (or other proton-accepting species) in solution:

$$NaOH \rightarrow Na^+ + OH^-$$

$$\text{sodium hydroxide} \rightarrow \text{sodium ion} + \text{hydroxide ion}$$

The OH^- is the effective base, since it can act as a proton acceptor:

$$OH^- + H^+ \rightarrow H_2O$$

$$\text{hydroxide ion} + \text{hydrogen ion} \rightarrow \text{water}$$

Potassium hydroxide is another widely used strong base. Calcium hydroxide is also a strong base, but it dissolves to only a slight extent in water, so that the concentration of hydroxide ions is always small.

QUESTION 5.1
Write balanced equations for:
(a) the reaction of nitric acid with water,
(b) the reaction of potassium hydroxide with water.

5.6 WEAK ACIDS AND WEAK BASES

Weak acids are only partially ionised in solution:

$$CH_3COOH + H_2O \rightleftharpoons H_3O^+ + CH_3COO^- \qquad (5.2)$$

ethanoic acid + water ⇌ hydronium ion + ethanoate anion

If read from left to right, the above equation shows ethanoic acid behaving as an acid. If read from right to left it shows the ethanoate ion acting as a base. Such complementary pairs of acids and bases are called **conjugates**: ethanoate ion is the conjugate base of ethanoic acid, while ethanoic acid is the conjugate acid of ethanoate ion.

Similarly, for weak bases, such as ammonia, NH_3:

$$NH_3 + H_2O \rightleftharpoons NH_4^+ + OH^- \qquad (5.3)$$

ammonia + water ⇌ ammonium ion + hydroxide ion

Reading from left to right, ammonia acts as a base, accepting a proton from the water molecule. Reading the equation the other way, the ammonium ion acts as an acid, donating a proton to the OH^-. Thus, ammonium ion is the conjugate acid of ammonia and ammonia is the conjugate base of ammonium ion.

Equations 5.2 and 5.3 also demonstrate that water can act as either a base or an acid. In Equation 5.2 water acts as a base, accepting a proton from the ethanoic acid, while in Equation 5.3 the water acts as an acid, donating a proton to the ammonia.

In Equation 5.1, the autoionisation equation, one water molecule acts as a base and the other as an acid. Water is thus the conjugate acid of the hydroxide ion (OH^-) and the conjugate base of the hydronium ion (H_3O^+).

5.7 K_a AND K_b

For the ionisation of a weak acid:

$$HA + H_2O \rightleftharpoons H_3O^+ + A^-$$

weak acid + water ⇌ hydronium ion + conjugate base

the equilibrium constant is given by:

$$K_{eq} = \frac{[H_3O^+][A^-]}{[HA][H_2O]}$$

However, as in the case of the equilibrium constant for the autoionisation of water, it is usual to omit from the expression the concentration of water itself, because this is large and (very nearly) constant. This gives us the dissociation constant of the acid:

$$K_a = \frac{[H_3O^+][A^-]}{[HA]}$$

Similarly, for a base:

$$B + H_2O \rightleftharpoons BH^+ + OH^-$$

$$\text{weak base} + \text{water} \rightleftharpoons \text{conjugate acid} + \text{hydroxide ion}$$

we obtain:

$$K_b = \frac{[OH^-][BH^+]}{[B]}$$

■ 5.8 RELATIONSHIP BETWEEN K_a AND K_b

For a weak acid the reaction:

$$HA + H_2O \rightleftharpoons H_3O^+ + A^-$$

$$\text{weak acid} + \text{water} \rightleftharpoons \text{hydronium ion} + \text{conjugate base}$$

has K_a given by:

$$K_a = \frac{[H_3O^+][A^-]}{[HA]}$$

The conjugate base of the acid can accept a proton from the water:

$$A^- + H_2O \rightleftharpoons HA + OH^-$$

and this reaction has K_b given by:

$$K_b = \frac{[HA][OH^-]}{[A^-]}$$

Multiplication of the dissociation constants for the weak acid and its conjugate base together gives:

$$K_aK_b = \frac{[H_3O^+][A^-]}{[HA]} \times \frac{[HA][OH^-]}{[A^-]}$$
$$= [H_3O^+][OH^-]$$
$$= K_w$$

$K_aK_b = K_w$

The two dissociation constants multiplied together thus give the autoionisation constant of water, K_w (section 5.2). Since K_w is a constant (= 10^{-14}) this means that as K_a increases K_b must get smaller. Therefore the stronger the acid is, the weaker its conjugate base becomes.

■ 5.9 pH, pOH, pK_w, pK_a AND pK_b

Because the values of $[H_3O^+]$ and the various kinds of K values span an enormous range (e.g. from 10^0 to 10^{-14} in the case of $[H_3O^+]$), it is customary to report these values in logarithmic terms. Thus, pH is defined as:

pH = $-\log_{10} [H_3O^+]$

$$\text{pH} = -\log_{10} [H_3O^+]$$

pOH is defined as:

$$\text{pOH} = -\log_{10} [\text{OH}]$$

Since

$$K_w = [H_3O^+][OH^-]$$

taking the negative values of the logarithms of the terms in this equation,

$$-\log_{10} K_w = -\log_{10} [H_3O^+] + (-\log_{10} [OH^-])$$

Since $-\log_{10} K_w = \text{p}K_w$, $-\log_{10} [H_3O^+] = \text{pH}$ and $-\log_{10} [OH^-] = \text{pOH}$, this gives us:

$$\text{p}K_w = \text{pH} + \text{pOH} = 14$$

The pH scale then runs from 0 to 14. Similarly, pK_w, pK_a and pK_b are defined as:

$$\text{p}K_w = -\log_{10} K_w = 14$$

$$\text{p}K_a = -\log_{10} K_a$$

$$\text{p}K_b = -\log_{10} K_b$$

Using these relationships, the following can be derived (see Appendix, Derivation 5.1):

pK_a + pK_b = pK_w

$$\text{p}K_a + \text{p}K_b = \text{p}K_w = 14$$

From what we have seen earlier, we can therefore write:

For a strong acid
pH = $-\log_{10} C$

(a) for a solution of a strong acid:

$$\text{pH} = -\log_{10} C$$

where C = concentration of the acid.

For a strong base
pH = pK_w + $\log_{10} C$

(b) for a solution of a strong base:

$$\text{pH} = \text{p}K_w + \log_{10} C$$

The derivations of these equations can be found in the Appendix, Derivations 5.2 and 5.3.

WORKED EXAMPLE 5.1

What would be the pH of a 0.05 mol dm^{-3} solution of the strong acid HCl?

ANSWER

$$\begin{aligned} pH &= -\log_{10} C \\ &= -\log_{10} 0.05 \\ &= 1.3 \end{aligned}$$

WORKED EXAMPLE 5.2

What would be the pH of a 0.02 mol dm^{-3} solution of the strong base NaOH?

ANSWER

$$\begin{aligned} pH &= pK_w + \log_{10} C \\ &= 14 + \log_{10} 0.02 \\ &= 14 + (-1.70) \\ &= 12.3 \end{aligned}$$

QUESTION 5.2

What would be the pH of a 0.1 mol dm^{-3} solution of HCl?

QUESTION 5.3

What would be the pH of a 0.015 mol dm^{-3} solution of NaOH?

5.10 SOLUTIONS OF WEAK ACIDS AND BASES

For weak acids, we have:

$$K_a = \frac{[H_3O^+][A^-]}{[HA]}$$

This leads to the equation (see Appendix, Derivation 5.4):

$$pH = \tfrac{1}{2}pK_a - \tfrac{1}{2}\log_{10} C$$

For a weak acid
$pH = \frac{1}{2}pK_a - \frac{1}{2}\log_{10} C$

For weak bases, we have:

$$K_b = \frac{[OH^-][BH^+]}{[B]}$$

By a similar line of reasoning (see Appendix, Derivation 5.5), we arrive at:

$$pH = pK_w - \tfrac{1}{2}pK_b + \tfrac{1}{2}\log_{10} C$$

For a weak base $pH = \frac{1}{2}pK_w + \frac{1}{2}pK_a + \frac{1}{2}\log_{10} C$

In many cases, pK_a values are listed for bases as well as for acids. If only pK_a values are available, the equation can be rewritten:

$$pH = \tfrac{1}{2}pK_w + \tfrac{1}{2}pK_a + \tfrac{1}{2}\log_{10} C$$

WORKED EXAMPLE 5.3

What would be the pH of a 0.05 mol dm^{-3} solution of the weak acid ethanoic acid, for which $pK_a = 4.75$?

ANSWER

We have:

$$\begin{aligned} pH &= \tfrac{1}{2}pK_a - \tfrac{1}{2}\log_{10} C \\ &= \tfrac{1}{2}(4.75) - \tfrac{1}{2}\log_{10} 0.05 \\ &= 2.375 - (-0.65) \\ &= 3.026 \end{aligned}$$

WORKED EXAMPLE 5.4

What would be the pH of a 0.1 mol dm^{-3} solution of the weak base ammonia, $pK_b = 4.75$?

ANSWER

We have:

$$\begin{aligned} pH &= pK_w - \tfrac{1}{2}pK_b + \tfrac{1}{2}\log_{10} C \\ &= 14 - \tfrac{1}{2}(4.75) + \tfrac{1}{2}\log_{10} 0.1 \\ &= 14 - 2.375 + (-1.0) \\ &= 11.625 \end{aligned}$$

WORKED EXAMPLE 5.5

What would be the pH of a 0.015 mol dm^{-3} solution of the weak base ethylamine, $pK_a = 10.81$?

ANSWER

We have:

$$\begin{aligned} pH &= \tfrac{1}{2}pK_w + \tfrac{1}{2}pK_a + \tfrac{1}{2}\log_{10} C \\ &= \tfrac{1}{2}(14) + \tfrac{1}{2}(10.81) + \tfrac{1}{2}\log_{10} 0.015 \\ &= 7 + 5.405 + (-0.912) \\ &= 11.493 \end{aligned}$$

QUESTION 5.4

What would be the pH of a 0.025 mol dm^{-3} solution of lactic acid, $pK_a = 3.86$?

QUESTION 5.5

What would be the pH of a 0.005 mol dm^{-3} solution of methylamine, $pK_b = 3.34$?

QUESTION 5.6

What would be the pH of a 0.015 mol dm^{-3} solution of trimethylamine, $pK_a = 9.81$?

■ 5.11 SALTS AND SALT HYDROLYSIS

Salts are produced by the reaction between acids and bases. All salts are fully dissociated into ions in solution. For salts of strong acids with strong bases this means that their solution consists of the ions and water only, e.g.

$$Na^+Cl^- + H_2O \rightarrow Na^+(aq) + Cl^-(aq)$$

sodium chloride + water → hydrated sodium ion + hydrated chloride ion

In the case of salts involving weak acids, however, the anion of the salt is the conjugate base of the weak acid. On dissolution in water, this will undergo hydrolysis. Thus, sodium ethanoate dissolves in water to produce sodium cations and ethanoate anions. The latter undergo hydrolysis (i.e. they react with water):

$$CH_3COO^- + H_2O \rightleftharpoons CH_3COOH + OH^- \qquad (5.4)$$

ethanoate anion + water ⇌ ethanoic acid + hydroxide ion

The solution is therefore basic, with

$$K_b = \frac{[CH_3COOH][OH^-]}{[CH_3COO^-]} \qquad (5.5)$$

This equation allows us to calculate the pH of solutions of salts of weak acids (see Appendix, Derivation 5.6)

$$pH = pK_w - \tfrac{1}{2}pK_b + \tfrac{1}{2}\log_{10} C$$

Often, only pK_a values for the parent acid will be available. In this case,

For a salt of a strong base with a weak acid $pH = \tfrac{1}{2}pK_w + \tfrac{1}{2}pK_a + \tfrac{1}{2}\log_{10} C$

$$pH = \tfrac{1}{2}pK_w + \tfrac{1}{2}pK_a + \tfrac{1}{2}\log_{10} C$$

In the case of salts involving weak bases, however, the action of the salt is the conjugate acid of a weak base. On dissolution in water, this will undergo hydrolysis. Thus, ammonium chloride dissolves in water to produce ammonium cations and chloride anions. The former undergo hydrolysis:

$$NH_4^+ + 2H_2O \rightleftharpoons NH_4OH + H_3O^+ \qquad (5.6)$$

ammonium ion + water ⇌ ammonium hydroxide + hydronium ion

The solution is therefore acidic, with

$$K_a = \frac{[NH_4OH][H_3O^+]}{[NH_4^+]} \qquad (5.7)$$

This equation allows us to calculate the pH of solutions of salts of weak bases (see Appendix, Derivation 5.7):

For a salt of a weak base with a strong acid $pH = \tfrac{1}{2}pK_a - \tfrac{1}{2}\log_{10} C$

$$pH = \tfrac{1}{2}pK_a - \tfrac{1}{2}\log_{10} C$$

If only pK_b data are available, the value for the parent base of the salt must be used, and the equation becomes:

$$pH = \tfrac{1}{2}pK_w - \tfrac{1}{2}pK_b - \tfrac{1}{2}\log_{10} C$$

WORKED EXAMPLE 5.6

What would be the pH of a 0.1 mol dm^{-3} solution of sodium propanoate? pK_a for propanoic acid = 4.87.

ANSWER

We have:

$$\begin{aligned} pH &= \tfrac{1}{2}pK_w + \tfrac{1}{2}pK_a + \tfrac{1}{2}\log_{10} C \\ &= \tfrac{1}{2}(14) + \tfrac{1}{2}(4.87) + \tfrac{1}{2}\log_{10} 0.1 \\ &= 7 + 2.435 + (-0.5) \\ &= 8.935 \end{aligned}$$

WORKED EXAMPLE 5.7

What would be the pH of a 0.002 mol dm^{-3} solution of triethylammonium chloride. pK_a for the triethylammonium ion = 10.76.

ANSWER

We have:

$$\begin{aligned} pH &= \tfrac{1}{2}pK_a - \tfrac{1}{2}\log_{10} C \\ &= \tfrac{1}{2}(10.76) - \tfrac{1}{2}\log_{10} 0.002 \\ &= 5.38 - (-1.35) \\ &= 6.73 \end{aligned}$$

QUESTION 5.7

What would be the pH of a 0.02 mol dm^{-3} solution of sodium methanoate? pK_a for methanoic acid = 3.75.

QUESTION 5.8

What would be the pH of a 0.01 mol dm^{-3} solution of diethylammonium chloride? pK_a for the diethylammonium ion = 10.99.

■ 5.12 BUFFER SYSTEMS

Many biochemical reactions are dependent on the pH of the solution in which they operate. This is particularly true for reactions involving enzymes, since the activity of the enzyme depends on the correct folding of the protein chain, and this folding can be affected by changes in pH. It is important, therefore, that pH is tightly controlled both in the living cell and in the laboratory. This control is achieved by the use of buffer solutions.

A buffer solution is one which resists changes in pH brought about by addition of acid or base. Buffer solutions are therefore used when it is vital that the pH be kept constant.

A buffer solution consists of a weak acid (or base) together with a solution of one of the salts of the weak acid (or base) with a strong base (or acid). Typical buffer solutions are mixtures of ethanoic acid (weak acid) with sodium ethanoate (salt of ethanoic acid with the strong base sodium hydroxide). In such a solution, the state of ionisation of the various components can be represented by:
Ethanoic acid:

$$CH_3COOH + H_2O \rightleftharpoons CH_3COO^- + H_3O^+$$

$$\text{ethanoic acid} + \text{water} \rightleftharpoons \text{ethanoate anion} + \text{hydronium ion}$$

Sodium ethanoate:

$$CH_3COO^- + Na^+$$

In the above solution, the ionisation of the ethanoic acid is suppressed by the presence of the ethanoate ions from the salt.

Consider what happens when hydrogen ions (i.e. an acid) are added to the solution. The hydrogen ions tend to combine with the ethanoate ions to produce more undissociated ethanoic acid. The concentration of hydrogen ions in the solution therefore does not change.

When hydroxide ions (i.e. a base) are added to the solution, these tend to combine with the few hydrogen ions present. The removal of these ions causes the ethanoic acid to dissociate further, producing more hydrogen ions. The buffer solution therefore resists changes to its hydrogen ion concentration.

■ 5.13 CALCULATING THE PHS OF BUFFERS

As mentioned above, the presence of the ethanoate anion suppresses the ionisation of the ethanoic acid. Therefore, the concentration of ethanoic acid in the buffer is much the same as the concentration of ethanoic acid in the solution originally:

$$[CH_3COOH] = [\text{acid}]_o$$

where $[\text{acid}]_o$ is the concentration of acid added originally.

For the same reason, the concentration of ethanoate anion present is much the same as the concentration of the salt in the solution originally:

$$[CH_3COO^-] = [\text{salt}]_o$$

where $[\text{salt}]_o$ is the concentration of salt in the solution originally.

The pH of a buffer can be shown to be given by the Henderson–Hasselboch equation (see Appendix, Derivation 5.8):

$$\text{pH} = \text{p}K_a + \log_{10} \frac{[\text{unprotonated species}]}{[\text{protonated species}]}$$

WORKED EXAMPLE 5.8

A buffer solution is made by mixing 200 cm^3 0.15 mol dm^{-3} ethanoic acid with 100 cm^3 of 0.15 mol dm^{-3} sodium ethanoate. What will be the pH of the resulting solution? pK_a for ethanoic acid = 4.75.

ANSWER

It is important to realise that in mixing the two solutions, both have been diluted.

$$\text{Volume of final solution} = 300\ \text{cm}^3$$

$$\text{Concentration of ethanoic acid in this solution} = \frac{200}{300} \times 0.15 = 0.10\ \text{mol dm}^{-3}$$

$$\text{Concentration of sodium ethanoate in this solution} = \frac{100}{300} \times 0.15 = 0.05\ \text{mol dm}^{-3}$$

We can then apply the Henderson–Hasselbach equation:

$$\text{pH} = \text{p}K_a + \log_{10} \frac{[\text{unprotonated species}]}{[\text{protonated species}]}$$

$$= 4.75 + \log_{10} \frac{0.05}{0.10}$$

$$= 4.79 + (-0.30)$$
$$= 4.49$$

WORKED EXAMPLE 5.9

Tris is a weak base frequently used to prepare buffers for biochemical use. Its full name is *tris*(hydroxymethyl)aminomethane, $pK_a = 8.08$. A volume of 500 cm^3 of a buffer at pH 8.2 is required, and we have available a 0.2 mol dm^{-3} solution of Tris. What volume of 2.0 mol dm^{-3} hydrochloric acid must be added to what volume of Tris to achieve this?

ANSWER

The first step is to determine what ratio of salt to base is required. The appropriate form of the Henderson–Hasselbach equation is:

$$\text{pH} = \text{p}K_a + \log_{10} \frac{[\text{unprotonated species}]}{[\text{protonated species}]}$$

$$8.2 = 8.08 + \log_{10} \frac{[\text{unprotonated species}]}{[\text{protonated species}]}$$

$$\log_{10} \frac{[\text{unprotonated species}]}{[\text{protonated species}]} = 8.2 - 8.08 = 0.12$$

$$\frac{[\text{unprotonated species}]}{[\text{protonated species}]} = 1.32$$

Let the volume of 0.2 mol dm^{-3} Tris required = V cm^3

Therefore, volume of acid required = 500 − V cm^3

Concentration of Tris after dilution to 500 cm^3 will be $\frac{0.2 \times V}{500}$ mol dm^{-3}

This is the concentration of the unprotonated species.

Concentration of acid after dilution to 500 cm^3 = $\frac{2.0 \times (500 - V)}{500}$ mol dm^{-3}

Since each mole of acid added to the solution produces one mole of salt, this will be the concentration of salt in the final solution. The salt in this case is the protonated species. We therefore have:

$$\frac{0.2 \times V}{500} \div \frac{2.0 \times (500 - V)}{500} = 1.32$$

$$\frac{0.2 \times V}{500} \times \frac{500}{2.0 \times (500 - V)} = 1.32$$

$$\frac{100 \times V}{1000 \times (500 - V)} = 1.32$$

$$100 \times V = 1320 \times (500 - V)$$

$$= 660\,000 - 1320\ V$$

$$1420\ V = 660\,000$$

$$V = 464.8 \text{ cm}^3$$

Therefore, in order to prepare the required amount of buffer, 464.8 cm^3 of the 0.2 mol dm^{-3} Tris must be mixed with 35.2 cm^3 of 2.0 mol dm^{-3} hydrochloric acid.

QUESTION 5.9

A buffer solution is made by mixing 200 cm^3 0.2 mol dm^{-3} ethanoic acid with 300 cm^3 0.15 mol dm^{-3} sodium ethanoate. What will be the pH of the final solution? pK_a for ethanoic acid = 4.75.

QUESTION 5.10

It is required to make 1000 cm^3 of a pH 8.0 Tris buffer. The available solutions are a 0.5 mol dm^{-3} solution of Tris and a 1.0 mol dm^{-3} solution of hydrochloric acid. What volumes of the two solutions must be mixed to achieve this? pK_a for Tris = 8.08.

■ 5.14 INDICATORS

Certain dyes have the property that they change colour depending on whether they are in an acidic or a basic solution. Such dyes are called **indicators** and can be used to detect acids or bases. Table 5.1 lists some common indicators, shows the acid, neutral and base colours produced and the pH range over which the colour change takes place. For example, methyl orange is red below pH 2.1, orange between pH 2.1 and 4.4 and yellow above pH 4.4.

Table 5.1 Some common indicators and their reactions

Name	*Acid*	*Neutral*	*Base*	*pH range*
Methyl orange	Red	Orange	Yellow	2.1–4.4
Methyl red	Red	Orange	Yellow	4.1–6.3
Bromothymol blue	Yellow	Green	Blue	6.0–7.6
Cresol red	Yellow	Orange	Red	7.2–8.8
Phenolphthalein	Colourless	Pink	Red	8.3–10.0
Alizarin red	Yellow	Orange	Red	10.1–12.0

It can be seen that the indicators change colour over different ranges. Universal Indicator consists of a mixture of a number of dyes. This gives a continuous range of colour changes as pH changes from about pH 1 to pH 12 so that the pH of a solution can be estimated by comparing the colour produced by the solution with a set of standard colours printed on a chart.

■ 5.15 TITRATIONS

The concentration of acid in a solution can be determined by reacting the acid in a known volume of the solution with a solution of base of known concentration. An indicator is used to show when all the acid has reacted, and the volume of the base solution required to just make the indicator change colour is noted.

The concentration of a solution of a base can be determined similarly using an acid solution of known concentration.

WORKED EXAMPLE 5.10

A 25.00 cm^3 sample of hydrochloric acid required 31.60 cm^3 of a 0.1000 mol dm^{-3} solution of sodium hydroxide to neutralise it. What was the concentration of the hydrochloric acid?

ANSWER

The equation for the reaction between hydrochloric acid and sodium hydroxide is:

$$HCl + NaOH \rightarrow NaCl + H_2O$$

hydrochloric acid + sodium hydroxide → sodium chloride + water

This shows that one mole of hydrochloric acid reacts with one mole of sodium hydroxide.

Number of moles of sodium hydroxide in 31.60 cm^3 of 0.1000 mol dm^{-3} solution

$$= \frac{0.1000}{1000} \times 31.60 = 0.00316$$

From the equation for the reaction, this will react with 0.00316 moles of hydrochloric acid.

This number of moles of hydrochloric acid was present in 25.00 cm^3 of solution.

1 dm^3 of the solution must therefore contain $\frac{0.00316}{25.00} \times 1000$ moles = 0.1264 moles

Concentration of hydrochloric acid = 0.1264 mol dm^{-3}

WORKED EXAMPLE 5.11

25.00 cm^3 of a solution of ammonia required 21.30 cm^3 of a 0.2000 mol dm^{-3} solution of sulphuric acid to neutralise it. What was the concentration of the ammonia solution?

ANSWER

The equation for the reaction between ammonia and sulphuric acid is:

$$2NH_3 + H_2SO_4 \rightarrow (NH_4)_2SO_4$$

ammonia + sulphuric acid → ammonium sulphate

This equation shows that two moles of ammonia react with one mole of sulphuric acid.
Number of moles of sulphuric acid in 21.30 cm^3 of a 0.2000 mol dm^{-3} solution

$$= \frac{0.2000}{1000} \times 21.30 = 0.00426$$

From the reaction equation, this number of moles will react with twice as many moles of ammonia.

Number of moles of ammonia = 2 × 0.00426 = 0.00852

This number of moles was present in 25.00 cm^3

1 dm^3 must therefore contain $\frac{0.00852}{25.00} \times 1000$ moles = 0.3408 moles

Concentration of ammonia = 0.3408 mol dm^{-3}

QUESTION 5.11

A 25.00 cm^3 sample of an ethanoic acid solution required 18.70 cm^3 of a 0.1000 mol dm^{-3} solution of sodium hydroxide to neutralise it. What was the concentration of the ethanoic acid solution?

The reaction equation is:

$$CH_3COOH + NaOH \rightarrow CH_3COONa + H_2O$$

ethanoic acid + sodium hydroxide → sodium ethanoate + water

QUESTION 5.12

A 25.00 cm^3 sample of a solution of sodium hydroxide required 15.25 cm^3 of a 0.05 mol dm^{-3} solution of sulphuric acid to neutralise it. What was the concentration of the sodium hydroxide?

The reaction equation is:

$$2NaOH + H_2SO_4 \rightarrow Na_2SO_4 + 2H_2O$$

sodium hydroxide + sulphuric acid → sodium sulphate + water

■ SUMMARY

The level of acidity of a solution is determined by the concentration of hydrogen (hydronium) ions. This is measured on the pH scale. Concentrations of acids, bases and the salts of weak acids with strong bases or strong acids with weak bases all affect the pH of a solution. The use of buffers allows pH values to be maintained at a desired level both in the laboratory and in living organisms.

The concentration of an acid or base can be determined by titration, using an indicator.

■ SUGGESTED FURTHER READING

Lehninger, A.L., Nelson, D.L. and Cox, M.M. (1993) *Principles of Biochemistry*, Ch. 4, 2nd edn. Worth, New York.

Atkins, P.W. (1994) *Physical Chemistry*, Ch. 9, 5th edn. Oxford University Press.

■ END OF CHAPTER QUESTIONS

Question 5.13 (a) Write an equation for the reaction of the weak acid methanoic acid, HCOOH, with water.

(b) Write an equation for the reaction of the weak base triethylamine, $N(CH_3)_3$, with water.

Question 5.14 (a) Calculate the pH of a 0.15 mol dm^{-3} solution of hydrochloric acid, HCl.

(b) Calculate the pH of a 0.01 mol dm^{-3} solution of potassium hydroxide, KOH.

Question 5.15 (a) Calculate the pH of a 0.05 mol dm^{-3} solution of the weak acid phenol, $pK_a = 9.89$.

(b) Calculate the pH of a 0.1 mol dm^{-3} solution of the weak base triethylamine. pK_a for the triethylammonium ion = 10.76.

Question 5.16 (a) What would be the pH of a 0.25 mol dm^{-3} solution of the salt sodium propionate. pK_a for propanoic acid = 4.87.

(b) What would be the pH of a 0.025 mol dm^{-3} solution of the salt methylammonium chloride. pK_a for the methylammonium ion = 10.66.

Question 5.17 It is required to prepare 100 cm^3 of a buffer of pH 4.5. A 0.2 mol dm^{-3} solution of ethanoic acid is available. pK_a for ethanoic acid = 4.75. (a) How many moles of sodium ethanoate must be added to this solution to achieve the required pH? (b) How many grams is this?

Question 5.18 For an experiment, 500 cm^3 of a buffer of pH 7.8 is required. A 0.5 mol dm^{-3} solution of ammonia is prepared. What volume of this solution must be mixed with what volume of 2.0 mol dm^{-3} hydrochloric acid to achieve the required pH? pK_a for the ammonium ion = 9.25.

Question 5.19 In a titration, 25.00 cm^3 of a solution of ethanoic acid required 19.6 cm^3 of a 0.100 mol dm^{-3} sodium hydroxide to neutralise it. What was the concentration of the ethanoic acid?

CARBON COMPOUNDS

6

6.1 INTRODUCTION

Carbon is a unique element. It has the unusual property of forming strong covalent bonds to other carbon atoms. This underlies organic chemistry, which is based on the formation of chain and ring structures of linked carbon atoms. Many biomolecules are complex organic molecules that are present in all living organisms. A second group of carbon compounds comprises simple structures in which there is usually only one carbon atom. They are important components of the mineral world, the sea and the atmosphere. Organisms use these substances as a carbon resource. The two areas of carbon chemistry are linked in the natural environment by the carbon cycle.

6.2 SIMPLE MOLECULES CONTAINING CARBON

Fossil fuels are a major carbon resource

The carbon-containing remains of many animals and plants occur in vast quantities in the earth's crust. Fossil fuels are typical of these remains; they are composed principally of carbon with the proportion present depending on the previous history of the material. The remains of long-dead plants and animals form these fuels – peat, coal, oil or natural gas. They are extracted and burned to provide heat but carbon dioxide is also formed, escapes into the atmosphere and is used by plants in photosynthesis. The carbon-containing minerals chalk, limestone and marble are mainly calcium carbonate. They are derived from the shells of tiny marine organisms which lived and died in large numbers. The shells sank to form sediments in shallow seas. Over long periods of time, they were subjected to heat and pressure in the earth to give the huge sedimentary deposits which exist today.

A solution of carbon dioxide in rainwater is a weak acid

Carbon exerts its valency of four by forming covalent bonds. It is often bound to oxygen or hydrogen. In the atmosphere, carbon exists as carbon dioxide. The slight solubility of carbon dioxide in rainwater allows it to form a weak acid, carbonic acid, in a reversible reaction:

$$\text{carbon dioxide} + \text{water} \rightleftharpoons \text{carbonic acid}$$

$$CO_2 + H_2O \rightleftharpoons H_2CO_3$$

Carbonic acid only exists in dilute solution. It reverts easily to carbon dioxide and water on warming or evaporating the solution. The solution is acidic because the compound forms bicarbonate (hydrogencarbonate) and hydrogen ions in a reversible ionisation:

$$\text{carbonic acid} \rightleftharpoons \text{hydrogen ions} + \text{bicarbonate ions}$$

$$H_2CO_3 \rightleftharpoons H^+ + HCO_3^- \ (pK_1)$$

The bicarbonate ion itself can ionise further to give more hydrogen ions and carbonate ions:

$$\text{bicarbonate ions} \rightleftharpoons \text{hydrogen ions} + \text{carbonate ions}$$

$$HCO_3^- \rightleftharpoons H^+ + CO_3^{2-} \ (pK_2)$$

The extent of each ionisation can be expressed in terms of the dissociation constant for the reaction (Chapter 5). Since the first reaction takes place to a much greater extent than the second, pK_1 is larger than pK_2. The slight solubility in water and the subsequent formation of hydrogen ions is of great significance. It means that rainwater is a weak acid and when it flows over rocks containing insoluble calcium carbonate, $CaCO_3$, the mineral is dissolved. This process occurs on a large scale and it means that the essential plant nutrient calcium is made readily available to the biosphere. It is solubilised as calcium bicarbonate (calcium hydrogencarbonate) in the chemical reaction:

Calcium carbonate rocks are mobilised by solution in rainwater

$$\text{calcium carbonate} + \text{carbonic acid} \rightleftharpoons \text{calcium bicarbonate}$$

$$CaCO_3 + H_2CO_3 \rightleftharpoons Ca(HCO_3)_2$$

The importance of calcium in the growth of plants has long been recognised. Land used to raise arable crops is treated with slaked lime on a regular basis to maintain soil fertility. When lime is not applied every few years, the crop yields fall steadily to a low level. Slaked lime is a convenient soluble form of calcium. It is made by heating limestone to decompose it. The quicklime, or calcium oxide, formed is treated with water to give slaked lime or calcium hydroxide.

QUESTION 6.1

Give equations to illustrate the reactions that occur when carbon dioxide dissolves in rainwater.

Explain how the reactions are involved in the mobilisation of calcium and carbon from chalk or limestone minerals.

Deposits of coal, oil and calcium carbonate rocks in the earth have their origin in living organisms. Mobilisation of these materials by burning fossil fuels or by the dissolution of carbonate rocks makes the carbon available once more for the biosphere. The organisms taking it up can themselves undergo death and decay. This can then lead to the formation of sediments and eventually to new deposits of carbon-containing minerals. In this way a cycle has been completed and can be sustained by natural processes. This is called the **carbon cycle**. The amounts of carbon compounds in the cycle are

Carbon moves between the mineral world and the biosphere through the carbon cycle

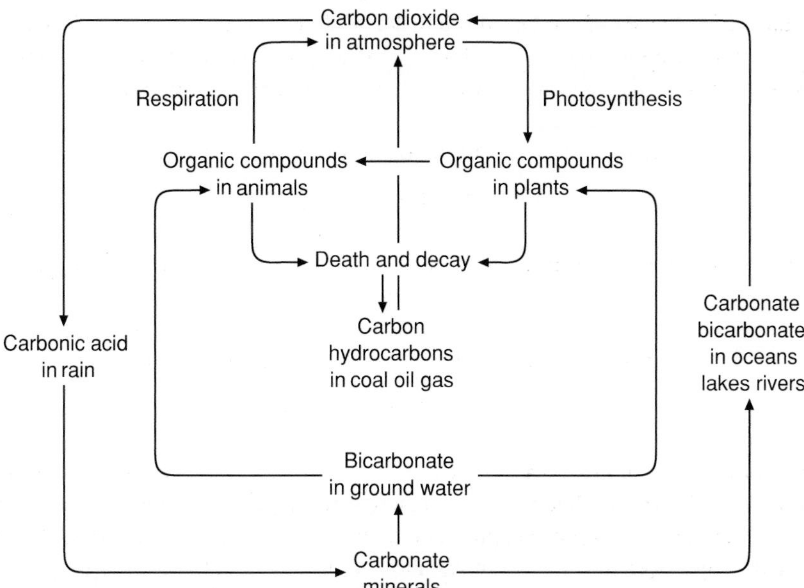

• **Figure 6.1** Schematic carbon cycle.

very large, allowing the cycle to be maintained in balance over a long period of time (Figure 6.1).

QUESTION 6.2

Describe three mechanisms in the carbon cycle that return carbon dioxide to the atmosphere.

■ 6.3 ORGANIC COMPOUNDS

Organic compounds are based on carbon atoms joined together into chains or rings with hydrogen, oxygen, nitrogen and a few other elements linked to them. The compounds often contain long carbon chains or rings. This is made possible because the atoms form strong bonds with one another. Most other elements form only weak bonds between their atoms. Thus the carbon–carbon bond energy is 348 kJ mol^{-1}. This is comparable with the energy of the bonds formed between carbon and elements such as oxygen, nitrogen and hydrogen: carbon–oxygen is 360 kJ mol^{-1} while carbon–nitrogen is 305 kJ mol^{-1} and carbon–hydrogen is 412 kJ mol^{-1}. High energies for the bonds between atoms within a molecule imply structural stability of the molecule. In some cases the high bond energy can provide a useful store of energy for an animal or a plant. The life processes of organisms require organic compounds to provide the diversity of structure and function needed by even the simplest microorganisms.

■ 6.4 ALKANES AND ALKYL GROUPS

Alkanes are hydrocarbons with a high proportion of hydrogen

Alkanes contain only carbon and hydrogen and are called **hydrocarbons**. Since carbon has a valency of four and hydrogen a valency of one, the simplest alkane hydrocarbon has four hydrogens joined to one carbon atom. This is methane, CH_4. When two carbon atoms are joined together then each has three hydrogen atoms attached to it to give ethane, C_2H_6. By adding further carbon atoms in turn, a series of alkane hydrocarbons is obtained.

Table 6.1 Names and formulae for alkane hydrocarbons

Name	*Molecular formula*	*Brief structural formula*	*Full structural formula*
Methane	CH_4	CH_4	H–C(H)(H)–H
Ethane	C_2H_6	CH_3CH_3	H–C(H)(H)–C(H)(H)–H
Propane	C_3H_8	$CH_3CH_2CH_3$	H–C(H)(H)–C(H)(H)–C(H)(H)–H
Butane	C_4H_{10}	$CH_3CH_2CH_2CH_3$	H–C(H)(H)–C(H)(H)–C(H)(H)–C(H)(H)–H

The next member is propane, C_3H_8. We can see that the formulae of methane and ethane differ by CH_2, as do the formulae of ethane and propane. A series of compounds where the members differ by CH_2 is called a **homologous series**. The alkanes form a homologous series. It is found that the members of such a series have similar chemical properties (Table 6.1). The alkanes are quite unreactive although they burn easily in air to give carbon dioxide and water with the production of heat. We burn alkanes as fuels. Natural gas is almost pure methane; it is burned in homes and power stations. Petrol is a mixture of several alkanes and is a convenient fuel for vehicles. All members of the alkane series obey the **general formula** C_nH_{2n+2}, where n is a whole number. Since the series contains a large number of compounds they have been made easy to recognise by giving the name of each one the same ending, or suffix, **-ane**. The first part of the name gives the number of carbon atoms in the molecule: meth = 1, eth = 2, prop = 3, but = 4. So when we are given the name of a compound, we can determine that it is an alkane and we know the number of carbon atoms in the molecule. Then using the general formula, we can write the formula for the compound. The simplest way of writing the formula for alkanes is called the **molecular formula**. Thus the molecular formula for ethane is C_2H_6. It is often useful to write the formula in a more descriptive way as a **structural formula**. Table 6.1 collects together the names, molecular and structural formulae for some members of the alkane series.

WORKED EXAMPLE 6.1

The molecular formulae (a) to (f) listed below are those of hydrocarbons:

(i) indicate which ones are alkanes,

(ii) name the alkanes you identify.

(a) C_3H_6 (b) C_4H_{10} (c) C_2H_6

(d) C_2H_4 (e) C_4H_6 (f) C_3H_8

ANSWER

(i) Alkanes have the general formula C_nH_{2n+2}, the examples that conform to this formula are (b), (c) and (f).

(ii) All the names end in -ane. The compound (b) has four carbon atoms which gives it the prefix but- (Table 6.1). Thus the name is butane; (c) and (f) are named in the same way to give ethane and propane.

QUESTION 6.3

The prefixes pent- and hex- correspond to five and six carbon atoms, respectively. Write the molecular and brief structural formulae for: (a) pentane; (b) hexane.

Although alkanes are of only limited importance in the life sciences, several features derived from them are of great value; one of these features is the **alkyl group**. When a hydrogen is removed from an alkane a group is left which could be joined to another molecule. The simplest alkyl group is methyl, CH_3—, which is formed by removing one hydrogen from methane, CH_4. The ethyl group, CH_3CH_2—, is formed from ethane, CH_3CH_3. Alkyl groups have a general formula, C_nH_{2n+1}, with n as a whole number. They are named in the same way as alkanes with the ending, or suffix, **-yl**, which is used by all alkyl groups. The first part of the name indicates the number of carbon atoms in the group. Alkyl groups do not have an independent existence, they must always be linked to other atoms or groups because each one has a spare valency. These groups are present in almost all biomolecules including amino acids, proteins, carbohydrates, lipids and nucleic acids. Often a number of similar substances differ only in the type of alkyl group present and the substances can be represented generally by replacing the formula of the alkyl group in each one by **R—**. When two alkyl groups are present in a structure, and they may not be the same, they can be represented as R and R′ or R^1 and R^2. Throughout the remainder of the book you will meet alkyl groups in all areas. Some examples are given in Table 6.2.

Alkyl groups are derived from alkanes and occur in most biomolecules

Table 6.2 Names and formulae for some alkyl groups

Name	*Brief structural formula (R =)*	*Full structural formula*
Methyl-	CH_3—	H | H—C— | H
Ethyl-	CH_3CH_2—	H H | | H—C—C— | | H H
Propyl-	$CH_3CH_2CH_2$—	H H H | | | H—C—C—C— | | | H H H

WORKED EXAMPLE 6.2

Give the names and full structural formulae for the alkyl groups formed from: (a) methane; (b) propane.

ANSWER

(a) Methane has the molecular formula CH_4. The alkyl group derived from it has one hydrogen less and thus has the full structural formula:

```
   H
   |
 H-C-
   |
   H
```

The name is made up of the prefix meth- for one carbon atom and the suffix -yl to give the name methyl.

(b) In the same way the alkyl derived from propane has the full structural formula:

```
   H H H
   | | |
 H-C-C-C-
   | | |
   H H H
```

and the name propyl.

QUESTION 6.4

Name the alkyl groups which are derived from the alkanes with molecular formula: (a) C_2H_6; (b) C_4H_{10}.

■ 6.5 ALKENES

A second homologous series of hydrocarbons exists and is distinguished by the presence of a carbon–carbon double bond. The first member of the series contains two carbon atoms since it is not possible to have a double bond with only one carbon atom. This is ethene, C_2H_4; it has the characteristic **-ene** suffix of the series and the two-carbon prefix eth-. The names and formulae of some alkenes are given in Table 6.3. From the table it is clear that they are named in the same way as alkanes.

Alkenes are much more reactive than alkanes, with the double bond providing a site at which addition can occur. This causes the double bond to be lost and gives a less reactive substance as the product. Thus ethene can undergo addition with water to give ethanol:

ethene + water → ethanol

```
   H                  H
   |                  |
 H-C    O-H         H-C-O-H
   ||  + |    →       |
 H-C    H           H-C-H
   |                  |
   H                  H
```

Table 6.3 Names and formulae for some alkenes

Name	*Molecular formula*	*Brief structural formula*	*Full structura formula*	*Alkyl group present (R =)*
Ethene	C_2H_4	$CH_2{=}CH_2$	H H \ / C=C / \ H H	H–
Propene	C_3H_6	$CH_3CH{=}CH_2$	H | H C–H \ / | C=C H / \ H H	CH_3–
Butene	C_4H_8	$CH_3CH_2CH{=}CH_2$	H H | | H C–C–H \ / | | C=C H H / \ H H	CH_3CH_2–

The addition of water to an alkene, hydration, followed by loss of hydrogen, dehydrogenation, to reform a different alkene is a key sequence in the citric acid cycle. Compounds, like alkenes, that can undergo addition are called **unsaturated** while those that do not take part in such reactions are termed **saturated**. Alkanes are saturated compounds. Ethene is a plant hormone implicated in senescence. It stimulates the ageing of flowers and the ripening of fruit. Within the molecule of an alkene it is the carbon–carbon double bond that provides a reactive site. An atom or a group of atoms within a molecule that is chemically reactive is called the **functional group**. The functional group in alkenes is the double bond.

Alkenes are unsaturated and undergo addition reactions

QUESTION 6.5

In the addition of water to ethene, a hydroxyl group adds to one of the carbon atoms forming part of the double bond, while a hydrogen atom adds to the second carbon of this bond. Draw the full structural formulae of the products formed when water adds to: (a) propene; (b) butene.

■ 6.6 ALCOHOLS

Alcohols differ from the alkane and alkene hydrocarbons in having a single oxygen atom in the molecule. This is present as the hydroxyl, —OH, group which is linked to an alkyl group. The second member of the alcohol series is ethanol, often simply called alcohol. It is familiar to us in a social and cultural context, where ethanol-based drinks are consumed today as they have been for many centuries. While ethanol is of relatively low toxicity when taken in limited amounts, other members of the series are dangerously poisonous, especially methanol. The alcohols are named in the same way as other homologous series with the characteristic **-ol** suffix denoting the presence of the hydroxyl functional group. Table 6.4 gives the names and formulae of some simple alcohols.

Alcohols, like other series of organic compounds, contain a functional group at which reactions occur

Table 6.4 Names and formulae for some alcohols

Name	*Molecular formula*	*Brief structural formula*	*Full structural formula*	*Alkyl group present (R =)*
Methanol	CH_4O	CH_3OH	H | H–C–OH | H	CH_3–
Ethanol	C_2H_6O	CH_3CH_2OH	H H | | H–C–C–OH | | H H	CH_3CH_2–
Propanol	C_3H_8O	$CH_3CH_2CH_2OH$	H H H | | | H–C–C–C–OH | | | H H H	$CH_3CH_2CH_2$–

QUESTION 6.6

Write the full structural formulae for the following alcohols (Tables 6.1 and 6.4): (a) propanol; (b) pentanol; (c) butanol.

QUESTION 6.7

Give the names of the alcohols represented by the molecular formulae: (a) C_2H_6O; (b) $C_4H_{10}O$; (c) CH_4O.

Alcohols occur in many different forms in plants and animals, especially as carbohydrates. In photosynthesis, plants use atmospheric carbon dioxide to form carbohydrates in the presence of sunlight. These are then stored as starch or cellulose and provide a vital food resource for animals. The carbohydrates are also used as a source of energy by plants. They are broken down in oxidative metabolism to give, ultimately, carbon dioxide and water. In the commercial world, ethanol is used as a fuel to produce heat. It burns with a blue flame and is efficiently converted to carbon dioxide and water with few partial combustion products and so is regarded as a 'clean' fuel:

$$\text{ethanol} + \text{oxygen} \rightarrow \text{carbon dioxide} + \text{water}$$

The partial oxidation of ethanol is brought about by microorganisms in air to form a carboxylic acid, ethanoic (acetic) acid. The microbes obtain energy from the reaction for their own metabolism:

$$\text{ethanol} + \text{oxygen} \rightarrow \text{ethanoic acid} + \text{water}$$

Alcohols undergo dehydration and oxidation

This reaction is responsible for the souring of wine when a partly consumed bottle is left open to the air for a few days. An alcohol can undergo loss of water on heating with a catalyst to give an alkene – this is **dehydration**. Thus ethanol give ethene:

$$\text{ethanol} \rightarrow \text{ethene} + \text{water}$$

This is the reverse of the reaction discussed earlier in which ethene underwent the addition of water (hydration) to form ethanol (section 6.5). Hydration and dehydration are important reactions in the citric acid cycle. Simple alcohols contain one hydroxyl group but it is possible for two or more to be present when the alcohol has several carbon atoms. Sugars often contain three, four or five hydroxyl groups while the long-chain storage carbohydrates starch and glycogen contain a large number of hydroxyl groups (section 7.10).

QUESTION 6.8

Suggest the products that are formed in the following reactions of alcohols:

(a) complete combustion of methanol in air,

(b) microbiological oxidation of ethanol in air,

(c) dehydration of propanol.

ADVANCED TOPIC BOX: ISOMERISM

Within the molecule of an organic or biological compound the carbon atoms or the functional group can often be arranged in more than one way. This means that two or more compounds with the same molecular formula can exist. This situation is called **isomerism**. Each of the different compounds is an **isomer**. Some molecular formulae can lead to just two isomers while others can form the basis of a large number of isomers.

When a functional group occurs in a molecule, it may be present on different carbon atoms; this leads to **structural isomerism**. In the case of the alcohol propanol, with molecular formula C_3H_8O, it is possible to place the hydroxyl group on either one of two different carbon atoms to give the structural formulae:

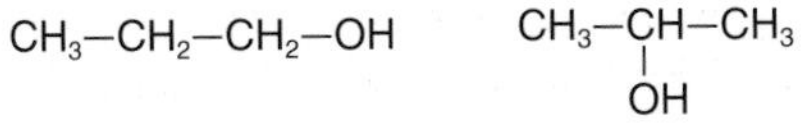

$$CH_3{-}CH_2{-}CH_2{-}OH \qquad CH_3{-}\underset{\displaystyle OH}{\underset{|}{CH}}{-}CH_3$$

These two compounds, propan-1-ol and propan-2-ol, are structural isomers with different chemical and physical properties.

Isomerism may be present even in molecules with the same structural formula. This involves different orientations in space of the groups within a molecule, and is called **stereoisomerism**. It is of major significance in the living world.

■ 6.7 THIOLS

Thiols are closely related to alcohols and contain a sulphur atom in place of the oxygen atom. The simplest thiol, methanethiol, CH_3SH, is a volatile, evil-smelling oil of negligible significance. Biologically important thiols are usually large molecules. The amino acid cysteine (Figure 6.2) contains the thiol group. When it is part of a protein structure it assumes significance in that it is easily oxidised. If two cysteine residues lie side by side in different peptide chains, or in the same chain, the thiol groups can form a sulphur–sulphur, —S—S—, covalent bond by oxidation:

Thiols are easily oxidised, they are biochemical intermediates

• **Figure 6.2** Cysteine

$$H_3\overset{+}{N}{-}\overset{\displaystyle SH \atop {| \atop {\displaystyle CH_2 \atop |}}}{C}{-}COO^- \atop \quad\;\; {| \atop H}$$

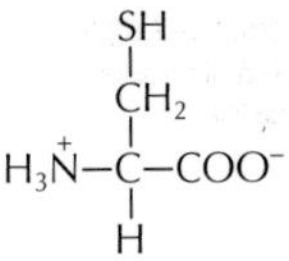

$$\text{cysteine} + \text{cysteine} \rightarrow \text{cystine} + \text{hydrogen}$$

This cystine bridge contributes to the tertiary structure of proteins. The cysteine residues are sufficiently reactive for oxygen in the air to oxidise them. The structural protein of hair, keratin, is rich in cysteine allowing large numbers of cystine bridges to be formed. These contribute to the straight or curly nature of hair. In 'permanent waving', the bridges can be broken by reduction and allowed to reform after the hair has been shaped.

QUESTION 6.9

2-Mercaptoethanol reduces cystine, $H_3N^+CHCH_2{-}S{-}S{-}CH_2CHNH_3^+$ (each CH bearing a COO^- group), to a thiol.

Write the brief structural formula of the thiol obtained.

The enzyme ribonuclease contains four cystine bonds between different parts of the molecule which contribute to its stability.

The acetyl transfer agent, Coenzyme A, shortened to CoA, is at the heart of pyruvate and thus carbohydrate oxidation. It is made up of adenosine triphosphate, the vitamin pantothenic acid and β-mercaptoethylamine:

ATP — Pantothenic acid — β-Mercaptoethylamine

A free thiol group in CoA is the active part of the molecule. It functions as an acyl transfer agent. Pyruvate, derived from sugars by glycolysis, transfers an acetyl group to CoA to form acetylCoA. This intermediate compound is an ester (sections 6.10 and 9.4). The acetyl group is then transferred to oxaloacetate giving citrate. The sequence is:

$$\text{pyruvate} + \text{CoA} \rightarrow \text{acetylCoA} + \text{carbon dioxide}$$

This reaction is mediated by the coenzyme nicotinamide adenine dinucleotide, NAD. The acetyl group is then transferred into the citric acid cycle:

$$\text{oxaloacetate} + \text{acetylCoA} \rightarrow \text{citrate} + \text{CoA}$$

■ 6.8 ALDEHYDES AND KETONES

Aldehydes are carbonyl compounds; they are readily oxidised or reduced

Oxidation of alcohols in air gives mainly carboxylic acid (section 6.6) but other products are formed in the reaction; these include aldehydes and ketones. The compounds contain a single oxygen atom but proportionately less hydrogen than alcohols. The characteristic aldehyde name ending is **-al**, and corresponds to the —CHO group. This group can only occur on the end carbon atom of a molecule. Ketones have names which end in **-one**, which denotes the —CO— group. Ketones complement aldehydes in that the ketone group can occur only in the middle of a carbon chain. Both aldehydes and ketones contain the **carbonyl** group, C=O, with a carbon–oxygen double bond. The names and formulae of some aldehydes and ketones are given in Table 6.5. From the table it can be seen that ethanal and propanal are aldehydes since each has the —CHO group and a name ending in -al. When the alkyl group in the brief structural formula is replaced by R— the

Table 6.5 Names and formulae for some aldehydes and ketones

Name	*Molecular formula*	*Brief structural formula*	*Full structural formula*	*Alkyl group present (R =)*
Methanal	CH_2O	HCHO	O ∥ H–C \\ H	H–
Ethanal	C_2H_4O	CH_3CHO	H O │ ∥ H–C–C │ \\ H H	CH_3–
Propanal	C_3H_6O	CH_3CH_2CHO	H H O │ │ ∥ H–C–C–C │ │ \\ H H H	CH_3CH_2–
Propanone	C_3H_6O	CH_3COCH_3	H O H │ ∥ │ H–C–C–C–H │ │ H H	CH_3–

general formula for aldehydes, RCHO, is obtained. Propanone has the —CO— group with the name ending -one and is a ketone. The general formula for ketones, RCOR, can be written by replacing the alkyl groups with R.

WORKED EXAMPLE 6.3

Assign each of the formulae listed as either an aldehyde or a ketone:

(a) $HOCH_2CHOHCHOHCHOHCHO$;

(b) $^-OOCCOCH_2CH_2COO^-$.

ANSWER

Write the full structural formula for each, checking that every carbon atom maintains a valency of four. Then identify either the aldehyde —CHO or ketone —CO— functional groups.

(a) structural formula: aldehyde group present,

```
      H   OH OH OH
      |   |  |  |      =O
HO—C—C—C—C—C
      |   |  |  |      \H
      H   H  H  H
```

(b) structural formula: ketone group present.

```
        O  H
O       ∥  |     O
  \C—C—C—C//
-O/        |    \O-
           H
```

QUESTION 6.10

Consider the list of brief structural formulae for carbonyl compounds. Identify each as either an aldehyde or a ketone.

(a) CH_3CH_2CHO (b) $CH_3CH_2COCH_3$

(c) $HOCH_2CH_2CHO$ (d) RCHO

(e) CH_3COCH_3 (f) $^-OOCCOCH_2COO^-$

• **Figure 6.3**
D-Glyceraldehyde

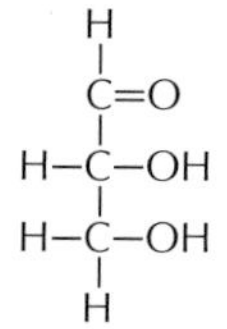

$H-C(=O)-C(H)(OH)-C(H)(OH)-H$

• **Figure 6.4**
Dihydroxyacetone

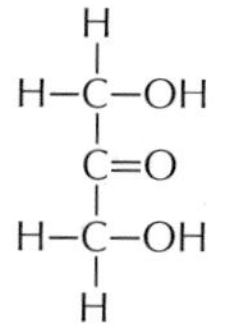

$H-C(H)(OH)-C(=O)-C(H)(OH)-H$

These compounds usually undergo reactions at the carbonyl group which are addition reactions (Chapter 7). These can lead to oxidation or reduction. It has been mentioned that simple sugars, monosaccharides, are alcohols (section 6.6) but they also contain an aldehyde or ketone group. This is typical of biomolecules which often have more than one type of functional group present. The simple sugars D-glyceraldehyde (Figure 6.3) and dihydroxyacetone (Figure 6.4) each have two hydroxyl groups and the carbonyl group.

■ 6.9 CARBOXYLIC ACIDS

When the hydroxyl, —OH, group is attached to the carbon of a carbonyl, —CO—, group, a carboxylic acid group is formed. It shows quite different properties from aldehydes, ketones or alcohols. This is because the —OH and —CO— groups interact to make the hydrogen atom acidic, giving the typical properties of an acid, such as sour taste. Carboxylic acids occur widely in the natural world and assume great importance in structural and metabolic substances. Thus all of the reactions in the citric acid cycle involve carboxylic acids. The names and formulae for some carboxylic acids are given in Table 6.6. It can be seen that the name of each has the suffix **-oic acid** together with a prefix showing the number of carbon atoms. Carboxylic acids which are important in the living world often carry the names they received when first recognised or prepared. These common names continue in use today and so we will use systematic or common names as appropriate. Both names are given in Table 6.7.

QUESTION 6.11
Give the systematic names for the brief structural formulae of the carboxylic acids: (a) CH_3CH_2COOH; (b) HCOOH; (c) HOOCCOOH.

Carboxylic acids can transfer a proton to a base or to water

Carboxylic acids usually dissolve in water and in doing so undergo ionisation, as shown for ethanoic (acetic) acid:

$$\text{ethanoic acid} + \text{water} \rightleftharpoons \text{ethanoate ion} + \text{hydroxonium ion}$$

The hydrogen ion lost by the acid is always taken up by a base with a lone pair of electrons. In this case, water is acting as a base. Ethanoic acid, like many carboxylic acids, is

Table 6.6 Names, formulae and pK_a values for some simple carboxylic acids

Name (common name)	*Brief structural formula*	*Full structural formula*	*Alkyl group present*	*pK_a*
Methanoic acid (formic acid)	HCOOH	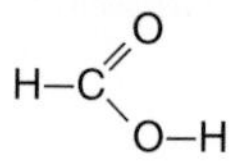 $H-C(=O)-O-H$	H–	3.85
Ethanoic acid (acetic acid)	CH_3COOH	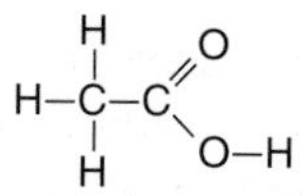 $H-C(H)(H)-C(=O)-O-H$	CH_3–	4.72
Propanoic acid	CH_3CH_2COOH	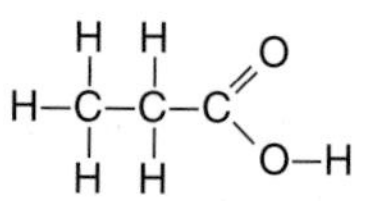 $H-C(H)(H)-C(H)(H)-C(=O)-O-H$	CH_3CH_2–	4.81

Table 6.7 Systematic names, common names and structural formulae for some carboxylic acids of biological significance

Systematic name	*Common name*	*Brief structural formula*
Methanoic acid	Formic acid	HCOOH
Ethanoic acid	Acetic acid	CH_3COOH
Butanoic acid	Butyric acid	$CH_3CH_2CH_2COOH$
Hexadecanoic acid	Palmitic acid	$CH_3(CH_2)_{14}COOH$
Octadecanoic acid	Stearic acid	$CH_3(CH_2)_{16}COOH$
cis,cis,-9,12-Octadecadienoic acid	Linoleic acid	$CH_3(CH_2)_4CH{=}CHCH_2CH{=}CH(CH_2)_7COOH$
Ethanedioic acid	Oxalic acid	HOOCCOOH
Propanedioic acid	Malonic acid	$HOOCCH_2COOH$
Butanedioic acid	Succinic acid	$HOOCCH_2CH_2COOH$
Butenedioic acid	Fumaric acid	$HOOCCH{=}CHCOOH$
Pentanedioic acid	Glutaric acid	$HOOCCH_2CH_2CH_2COOH$
3-Carboxypentan-3-oldioic acid	Citric acid	OH (on central C) $HOOCCH_2CCH_2COOH$ COOH (on central C)
Pentan-2-onedioic acid	2-Ketoglutaric acid	O (double bond, on C next to COOH) $HOOCCH_2CH_2CCOOH$

only partly ionised in water. The term **weak acid** is used to describe this situation where the solution in water contains molecules of the acid, as well as the ions formed from it. The extent of ionisation of a carboxylic acid is measured by the pK_a value (section 5.9). A low pK_a value corresponds to an acid that is highly ionised while a high value means that few ions are formed. The names of some biologically important carboxylic acids are listed in Table 6.7.

QUESTION 6.12

Write an equation, using brief structural formulae, to show the ionisation in water of butanoic acid.

Carboxylic acids can be converted to amides and esters

Within the cell, carboxylic acids take part in several important reactions including amidation, decarboxylation and acyl substitution (esterification). Amidation is the conversion of the acid to an **amide** by reaction with an amine. For ethanoic (acetic) acid and ethylamine the reaction is:

ethanoic acid + ethylamine → *N*-ethylethanamide + water

$$CH_3COOH + CH_3CH_2NH_2 \rightarrow CH_3CONHCH_2CH_3 + H_2O$$

This reaction is the basis for formation of the peptide bond, —CONH—. It usually takes place between two amino acids to form a dipeptide that has a carboxyl group at one end of the molecule and an amide group at the other. This makes it capable of reacting further with additional amino acids to give, eventually, a long peptide chain. Such chains form the basis of proteins.

Decarboxylation is the loss of carbon dioxide from a carboxylic acid

Decarboxylation is the loss of carbon dioxide from a carboxylic acid to form an alkane. For simple acids, the conditions required are more aggressive than can be achieved in the living cell. But keto-acids and di- or tricarboxylic acids are more easily decarboxylated; such reactions occur in the citric acid cycle. Thus, the isocitrate ion is converted to the 2-ketoglutarate ion by way of the easily decarboxylated intermediate, oxalosuccinate ion:

$$\text{isocitrate} \rightarrow \text{2-ketoglutarate} + \text{carbon dioxide}$$

The reaction is mediated by enzyme catalysts and overall is more complex than suggested here.

Esters are formed by the acyl substitution of carboxylic acids. For energetic reasons within the cell, the reaction does not occur directly between a carboxylic acid and an alcohol but proceeds by way of a thio ester which is formed from a thiol (section 6.7) and a carboxylic acid (section 7.2). An ester has the general formula RCOOR′; an example is ethyl ethanoate, $CH_3COOCH_2CH_3$.

A carboxylic acid molecule contains both hydroxyl, —OH, and carbonyl, —CO—, groups and so it is able to form hydrogen bonds (section 3.4) with other carboxylic acid molecules. For ethanoic acid the reaction is:

$$H_3C{-}C(=O){-}O^{\delta-}{-}H^{\delta+}\cdots{}^{\delta-}O{=}C(OH){-}CH_3$$

It can be seen from this diagram that one carboxylic acid has an unused hydroxyl group while the other has a carbonyl group available. These groups now combine to form a second hydrogen bond linking the two molecules together strongly:

$$H_3C{-}C(={}^{\delta-}O\cdots{}^{\delta+}H{-}{}^{\delta-}O{-})(-O{-}H\cdots O{=})C{-}CH_3$$

In the same way, hydrogen bonds can be formed between carboxylic acids and water, alcohols, thiols, amines or carbonyl groups. These bonds are of structural significance in biomolecules.

QUESTION 6.13

Two types of hydrogen bonding are possible between ethanoic acid and water. Draw brief structural formulae to show these hydrogen bonds.

Table 6.8 Names and brief structural formulae for some amines

Name	*Brief structural formula*	*Primary, secondary or tertiary amine*
Methylamine	CH_3NH_2	Primary
Ethylamine	$CH_3CH_2NH_2$	Primary
Propylamine	$CH_3CH_2CH_2NH_2$	Primary
Dimethylamine	CH_3NHCH_3 or $(CH_3)_2NH$	Secondary
Diethylamine	$CH_3CH_2NHCH_2CH_3$ or $(CH_3CH_2)_2NH$	Secondary
Trimethylamine	$CH_3N(CH_3)CH_3$ (N bonded to CH_3 below) or $(CH_3)_3N$	Tertiary

■ 6.10 AMINES

Amines are derived from the simple nitrogen-containing compound ammonia, NH_3. Replacement of a hydrogen atom by an alkyl group forms an amine. The first amine in the homologous series is methylamine or methanamine, CH_3NH_2. In order to name the compounds, the **-amine** suffix is added to the name of the alkyl group as a prefix (methylamine), or alternatively it is added to a prefix formed by removing -e from the parent alkane (methanamine). The second amine in the series is ethylamine or ethanamine, $CH_3CH_2NH_2$, with an ethyl group linked to nitrogen. Several amines are listed in Table 6.8.

Since methylamine has two hydrogen atoms bound to nitrogen, other amines can be obtained by replacing one, or both, of these hydrogens with alkyl groups. When two methyl groups are present, the amine is dimethylamine, $(CH_3)_2NH$, while three methyl groups linked to nitrogen gives trimethylamine, $(CH_3)_3N$. The methyl groups can be replaced by other alkyl groups which can be the same or different when linked to the same nitrogen. In this way, we can identify three types of amines with general formulae: RNH_2, R_2NH and R_3N. These are called **primary**, **secondary** and **tertiary** amines, respectively. They are easily distinguished by counting the number of hydrogen atoms linked to nitrogen. When two are present, the compound is a primary amine. One hydrogen atom bound to nitrogen is indicative of a secondary amine. With three alkyl groups and a nitrogen atom having no hydrogens joined to it, the compound is a tertiary amine. Some examples are listed in Table 6.8. Amines of all types are important in living organisms.

QUESTION 6.14

Six amines are listed below as brief structural formulae. Consider each one and describe it as a primary, secondary or tertiary amine.

(a) $CH_3CH_2CH_2NH_2$ (b) $CH_3CH_2NHCH_3$

(c) H_2NCH_2COOH (d) $(CH_3)_3N$

(e) $HOCH_2CH_2NH_2$ (f) $CH_3CH_2NHCH_2CH_3$

One particularly important group of amines contain both the amine and the carboxyl groups. These are the amino acids. They usually have the two functional groups linked to the same carbon atom. The first member of the amino acid series is aminoethanoic acid, usually called glycine, H_2NCH_2COOH.

Amines react with acids to form salts

Amines are weak bases and so react with acids to form salts. It is the lone pair of electrons on nitrogen that is active in accepting the proton, H^+, from an acid. The process can be illustrated by comparing the reactions of ammonia and methylamine with hydrochloric acid:

$$\text{ammonia} + \text{hydrochloric acid} \rightarrow \text{ammonium chloride}$$

$$NH_3 + HCl \rightarrow NH_4^+Cl^-$$

This reaction is probably familiar; it shows that the ammonium ion is formed when ammonia receives a proton. The salt produced is ammonium chloride.
Methylamine behaves in the same way to give a methylammonium ion:

$$\text{methylamine} + \text{hydrochloric acid} \rightarrow \text{methylammonium chloride}$$

$$CH_3NH_2 + HCl \rightarrow CH_3NH_3^+Cl^-$$

QUESTION 6.15

Write the structural formulae for the ammonium ions formed when the amines given below react with the hydrogen ion, H^+.

(a) $CH_3CH_2NH_2$ (b) $(CH_3)_3N$
(c) $HOCH_2CH_2NH_2$ (d) CH_3NHCH_3

Amines have a highly electronegative atom, nitrogen, linked to hydrogen and so show the criteria for hydrogen bonding (section 3.4). They readily form these bonds with one another:

$$\overset{\delta-}{N}-\overset{\delta+}{H}\cdots\overset{\delta-}{N}$$

Carboxylic acids and amines can each form hydrogen bonds to a range of other compounds

Hydrogen bonding to other groups such as a carbonyl oxygen, an alcohol hydrogen or oxygen and a carbonyl hydrogen or oxygen is significant. It plays a major role in maintaining the folded structure of proteins and the double helix form of nucleic acids.

R–NH–C(=O)–R

An amide

O, N, H

dashed line: delocalised electrons

ADVANCED TOPIC BOX: AMIDES AND THE PEPTIDE BOND

An amine can react with a carboxylic acid to form an **amide** (see left). The amide formed is not a base and is much less reactive than the corresponding amine. When amino acids are joined together in peptide synthesis, amide bonds are formed.

In an amide the electrons from the carbonyl oxygen are delocalised and are shared with the amide bond (see left). This means that the bond around the amide has partial double bond characteristics. This affects the way in which a polypeptide chain can fold.

■ SUMMARY

Carbon is distributed in the earth as carbonate minerals, as atmospheric carbon dioxide and as fossil fuels. Carbon moves into the biosphere when carbon dioxide is taken up by green plants. Animals rely on plants in a direct or indirect way for the element. The death and decay of plants or animals returns carbon to the non-living world. The tendency of carbon atoms to link together in chains or rings provides the basis for organic chemistry. Alkanes and alkenes each form series of hydrocarbon compounds. When oxygen is present in an organic compound then alcohols, aldehydes, ketones and carboxylic acids can occur. Each of these types of compound form a homologous series and each has a functional group at which reactions occur. In such a series the compounds have similar properties and names and differ from their neighbours by $-CH_2-$. The incorporation of nitrogen into organic compounds leads to the amine and amide series while the presence of sulphur gives thiols. All of these compounds are intimately involved in the structure and function of organisms.

■ END OF CHAPTER QUESTIONS

Question 6.16 How are (a) coal and (b) chalk formed in the earth's crust?

Question 6.17 Outline the mechanisms by which carbon in:

(a) calcium carbonate (b) carbon dioxide

move from the non-living world to living organisms in the carbon cycle.

Question 6.18 The alkene hydrocarbons:

(a) ethene (b) butene

can be formed by dehydration of the corresponding alcohols. Give the brief structural formulae of the alcohols required.

Question 6.19 The amino acid cysteine undergoes oxidation under mild conditions to form a compound with a disulphide link:

(a) give the brief structural formula of this compound,
(b) what is the significance of this linkage in protein structures?

Question 6.20 The alcohols listed can be oxidised to either an aldehyde or a ketone:

(i) ethanol, CH_3CH_2OH
(ii) 2-butanol, $CH_3CH_2CHOHCH_3$
(iii) 1-butanol, $CH_3CH_2CH_2CH_2OH$

For each product formed:

(a) identify it as an aldehyde or a ketone,
(b) name the compound,
(c) write the brief structural formula.

Question 6.21 Carboxylic acids lose carbon dioxide by decarboxylation. What products are obtained by decarboxylation of the following acids? Give the name or structural formula:

(a) propanoic acid, CH_3CH_2COOH
(b) 2-ketobutanoic acid (oxaloacetic acid), $CH_3CH_2COCOOH$

Question 6.22 Draw diagrams to show the following types of hydrogen bonding between ethanoic acid and ethanol:

(a) carboxyl-oxygen to hydrogen,
(b) carboxyl-hydroxyl to oxygen.

Question 6.23 Write word and symbol equations to show the reaction between methylamine and hydrochloric acid in water.

7

LIPIDS, SUGARS, AND LINKAGES BETWEEN REACTIVE GROUPS

■ 7.1 INTRODUCTION

The simple functional groups that have been described in Chapter 6 are components of many of the important classes of biological compounds. The functional groups on biomolecules provide a means of synthesising the complex macromolecules found in living cells, such as fats and oils, as well as a place to begin the metabolic breakdown of such compounds. One important class of biomolecules comprises sugars and the complex carbohydrates that result from the linkages between sugars. Sugars play a major part in the biosphere, contributing a useful easily metabolised food store and forming one of the major structural components of animal and plant tissue. Fats and oils constitute another major group of biomolecules which functions in the living cell as an energy store and provides the precursors of biological membranes.

■ 7.2 FATTY ACIDS

Typical organic acids include the simple carboxylic acids (Figure 7.1) which consist of an alkyl chain with a carboxyl group at one end. Long-chain carboxylic acids are known as fatty acids, as they are a major component of fats and oils. The alkyl chain can vary in length up to 30 carbons and can contain alkene functional groups within its structure. Carboxylic acids can be part of a biological molecule such as citric acid which contains other functional groups.

CH_2-COOH
$HO-CH-COOH$
CH_2-COOH

Citric acid contains three carboxyl groups and a hydroxyl group

Carboxylic acids are named by reference to the parent alkane from which they are derived. Thus, HCOOH has one carbon and is called methanoic acid and CH_3COOH contains two carbons and is called ethanoic acid (section 6.10). The carboxylic acids of particular interest to life scientists contain long alkyl chains. Long-chain carboxylic acids are also called fatty acids as they are one of the components of fats and oils. Most naturally occurring fatty acids vary in chain length between 12 and 20 carbons. The names associated with some fatty acids are shown in Table 7.1. It can be seen from Table 7.1 that fatty acids have two names, one of which is systematic (and is preferred) whilst the other is the commonly used name found in most biochemistry textbooks.

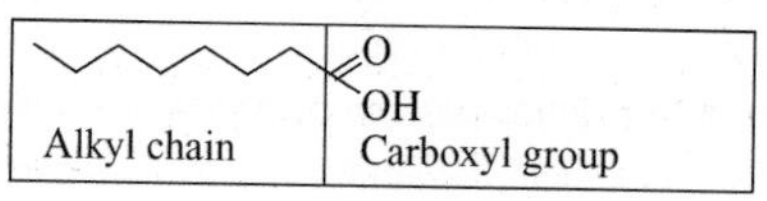

• **Figure 7.1** Structure of a carboxylic acid.

Table 7.1 The names of some saturated fatty acids and their melting points

Systematic acid name	*Trivial name*	*Number of carbon atoms*	*Melting point (°C)*
Dodecanoic	Lauric	12	44.8
Tetradecanoic	Myristic	14	54.4
Hexadecanoic	Palmitic	16	62.9
Octadecanoic	Stearic	18	70.1
Eicosanoic	Arachidic	20	76.1
Docosanoic	Behenic	22	80.0

Naming fatty acids is further complicated by the presence of alkene groups. Fatty acids containing no alkene groups are described as saturated fatty acids whilst those containing one or more alkene groups are described as unsaturated fatty acids. The alkene group in nearly every naturally occurring fatty acid is found in the *cis* configuration. *cis* unsaturated fatty acids cannot be synthesised by the human body and are essential components of the diet. Such fatty acids are often called essential fatty acids. Systematically, alkene groups are named by counting the carbons from the carboxyl end of the acid, the so-called delta (Δ) nomenclature. Some textbooks still refer to the position of alkene groups by reference to the methyl end of the alkyl chain (the ω notation).

Natural unsaturated fatty acids are normally *cis* isomers

WORKED EXAMPLE 7.1

Name the fatty acid shown in Figure 7.2.

• **Figure 7.2** The structure of an unsaturated fatty acid.

ANSWER

Number the carbons from the carboxyl end. This is already done in Figure 7.2. There are 18. This means that the fatty acid is octadec–oic acid (see Table 7.1).

Count how many alkene groups are present. In this case there is only one. The fatty acid is therefore octadecenoic acid. Two, three or four alkene groups would give octadecadienoic, octadecatrienoic and octadecatetraenoic acid, respectively.

Number the type and position of the alkene group. The alkene in Figure 7.2 is in the *cis* configuration and is 9 carbons from the carboxyl end and so the full name is *cis*-9-octadecenoic acid.

QUESTION 7.1

Draw the structure of *cis*-9,12-hexadecadienoic acid.

Table 7.2 The names and melting points of some unsaturated fatty acids

Name		*Melting point (°C)*		
Systematic	*Trivial*	cis	trans	*Shorthand*
cis-9-Tetradecenoic	Myristoleic	–4.0	18.5	$14{:}1\Delta^9$
cis-9-Hexadecenoic	Palmitoleic	0.5	32	$16{:}1\Delta^9$
cis-6-Octadecenoic	Petroselenic	29	54	$18{:}1\Delta^6$
cis-9-Octadecenoic	Oleic	16	45	$18{:}1\Delta^9$
cis-11-Octadecenoic	*cis*-Vaccenic	15	44	$18{:}1\Delta^{11}$
cis-11-Eicosenoic	Gondoic	24	—	$20{:}1\Delta^{11}$
cis-11-Docosenoic	Erucic	34	60	$22{:}1\Delta^{11}$

Number of double bonds ↘ 18:0 ↗ Length of chain

Meaning of shorthand way of writing fatty acid structures

The full names are unwieldy and so common names have retained their use. The fatty acid in Figure 7.2 is also called oleic acid. A shorthand way of writing the formula of unsaturated fatty acids is to write the chain length followed by the number of the double bond separated by a colon (chain:alkenes). Thus oleic acid would be 18:1 and stearic acid 18:0. In this nomenclature the position of the double bond is indicated by the way of naming the position followed by the number in superscript. Thus oleic acid is abbreviated to $18{:}1\Delta^9$. Table 7.2 shows the names of some commonly occurring unsaturated fatty acids.

cis unsaturation lowers fatty acid melting points

Table 7.1 shows that the melting point of fatty acids increases as the chain length increases. This is due to an increase in the ability of the alkyl chains to undergo van der Waals interactions (see Chapter 3) as the length of the alkyl chain increases. The situation is complicated by the presence of *cis* unsaturation. Figure 7.2 shows that the *cis* double bonds cause alkyl chains to kink (at an angle of 60°). This decreases the possibility of the sort of proximity required for van der Waals interactions and so decreases markedly the melting point of the fatty acid. Oleic acid (*cis* 18:1) has a melting point of 16°C compared with 70.1°C for stearic acid (18:0).

The carboxyl group contains both a **carbo**nyl and hydro**xyl** group and its chemical properties derive from this combination. Thus the carboxyl carbon is electropositive and represents a good site for nucleophilic attack whilst the hydroxyl proton can dissociate giving the carboxyl group its characteristic acidic property.

WORKED EXAMPLE 7.2

Explain which of the carboxylic acids given in Figure 7.3(a) will have the higher melting temperature.

(a) (i) (ii) O OH O OH

(b) (i) (ii) OH O O OH

• **Figure 7.3** Some fatty acids.

ANSWER

Count the length of the carbon chain. The first fatty acid has 14 carbons whilst the second has 16. The longest will have the higher melting temperature as there is a greater probability of van der Waals interactions. Then check the number of *cis* double bonds. The number of *cis* double bonds decreases the melting temperature as van der Waals interactions are lowered by kinking of the alkyl chain. Both fatty acids are fully saturated. Fatty acid (ii) will have the higher melting temperature.

QUESTION 7.2

Explain which of the carboxylic acids given in Figure 7.3(b) will have the higher melting temperature.

7.3 ESTERS

The most common ester is the product of the reaction between an organic acid and an organic base, such as an alcohol (although phosphate and thiol esters will be covered in Chapter 9). This reaction is shown in Figure 7.4. The process of **ester bond** formation is called esterification and the acid is said to be esterified. It should be noted that this is a condensation reaction as water is lost. The hydroxyl group in the water that is lost comes from the carboxyl group whilst the other water proton comes from the hydroxyl of the reacting alcohol. Esters are named by reference to the reacting groups. Thus the reaction between ethanoic acid and methanol yields methyl (from methanol) and ethanoate (from ethanoic acid) to give methyl ethanoate.

The ester bond

$$\mathbf{R{-}C(=O)OH} + HO{-}R' \rightleftharpoons \mathbf{R{-}C(=O)}{-}O{-}R' + \mathbf{H{-}O{-}}H$$

Carboxylic acid | Alcohol | Ester

• **Figure 7.4** The formation of an organic ester (acid shown in bold). Note that the oxygen in the ester bond linking the acid and alcohol derives from the alcohol.

WORKED EXAMPLE 7.3

Draw and name the ester formed from the reaction between propanoic acid and methanol.

ANSWER

Write out the formulae of the reactants: CH_3CH_2COOH and CH_3OH.

Reverse the alcohol to give $HOCH_3$.

Remove the OH from the propanoic acid and the H from the methanol to give CH_3CH_2O– and –OCH_3.

Join up to give $CH_3CH_2COOCH_3$.

Propanoic acid forms the ester and so becomes propanoate in the ester whilst the alcohol is methanol and becomes methyl in the ester to give the name methyl propanoate for the ester.

QUESTION 7.3

Draw and name the ester formed from the reaction between methanoic acid and ethanol.

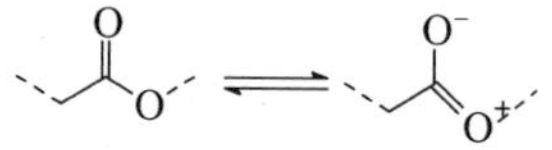

• **Figure 7.5** Resonance stabilisation of the carboxyl group.

Acetylsalicylic acid or aspirin

Ester bonds have a decreased polarity when compared to the carboxyl group as a result of resonance stabilisation of the ester bond (Figure 7.5). The electron-withdrawing effect of both ester oxygens makes the carboxyl carbon electropositive. It is therefore a good site for nucleophilic attack. Esterification of carboxyl groups also reduces polarity and therefore esterification can be used to aid the passage of drugs into the blood and from the blood to cells. Esterification is used to convert salicylic acid to ethanoylsalicylic acid (more commonly known as acetylsalicylic acid or aspirin). This compound is less irritating to the stomach than its parent compound. Esters are volatile compounds that form some of the many aromas that we can sense. These include the characteristic aromas of pineapples (ethyl butanoate) and pears (3-methyl-butyl ethanoate) amongst others.

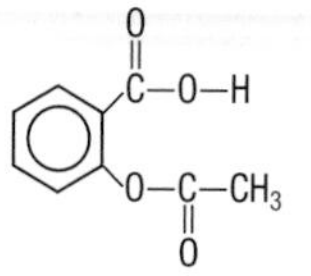

Waxes are esters of long chain alcohols and fatty acids

Esters of long-chain fatty acids and long-chain alcohols form a class of biomolecules known as waxes. The ester of hexadecanoic acid and hexadecanol, hexadecanyl hexadecanoate, comprises nearly all of the 'fat' in whale blubber. Waxes are highly non-polar molecules and occur in nature as natural water repellents as found in bird feathers. The hardness of the wax is dependent on the length of the two chains involved as well as the number of *cis* alkene bonds.

Esters (particularly thiol esters discussed in Chapters 6 and 9) are often used in biological reactions as intermediates in the synthesis of metabolites where a direct synthesis would not be possible at 37°C in aqueous media.

7.4 GLYCEROL ESTERS

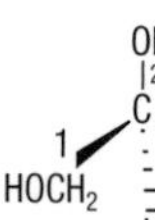

Propan-1,2,3-triol

Propan-1,2,3-triol is more commonly known as glycerol. The structure shows the presence of three hydroxyl groups. Each of the hydroxyl groups can be esterified. The product of esterification of one hydroxyl group by a fatty acid yields a monoacylglycerol or monoglyceride. A typical monoglyceride (see Figure 7.6) is glycerol monostearate (GMS), a commonly used emulsifier in foods. Two or three hydroxyl groups can be esterified to yield diacylglycerols (diglycerides) or triacylglycerols (triglycerides) respectively.

Triacylglycerols are the most common constituents of fats and oils. Fats differ from oils in being solid rather than liquid at room temperature. This is because the fatty acyl chains are more saturated. The fatty acid in olive oil, for example, contains predominantly *cis*-9-octadecenoic acid, whilst palm fat contains mainly dodecanoic acid. The fatty acids which are ester linked to glycerol may be identical or may be different. In naturally occurring fats and oils the different fatty acyl chains are distributed to give the maximum number of different possible mixtures. An additional complexity arises because the carbons in triglycerides are all different. At first glance the 1 and 3 positions in a triglyceride which contain non-identical fatty acyl chains look equivalent. However, this does not take into account the three-dimensional structure of the molecule (see margin). In the figure in the margin the alkyl chains ester bonded to the glycerol carbon 1 and 3 have been swopped. You can see that the hydrogen attached to glycerol carbon 2 has shifted

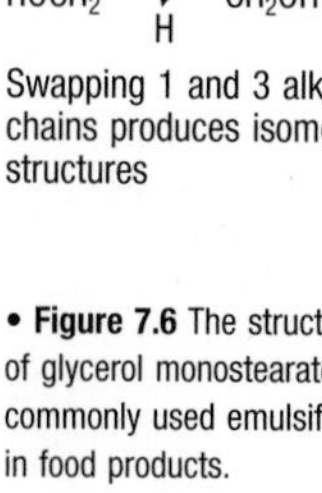

Swapping 1 and 3 alkyl chains produces isomeric structures

• **Figure 7.6** The structure of glycerol monostearate, a commonly used emulsifier in food products.

its position with respect to that of carbon. Consequently, enzymes that hydrolyse the ester linkages in triglycerides (lipases) can be specific to either the 1 or 3 carbon. The carbons are therefore numbered from the top in Fisher projection as 1, 2 or 3, respectively. The numbering is given an *sn* notation (*sn* is short for stereochemical notation).

WORKED EXAMPLE 7.4

Explain the difference between a diacylglycerol and a triacylglycerol.

ANSWER

Diacylglycerols are esters between two of the hydroxyl groups in glycerol and two fatty acids whilst triacylglycerols are the product of esterification of glycerol by three fatty acids.

QUESTION 7.4

Explain why glyceryl trioleate is an oil whilst glyceryl tristearate is a fat.

■ 7.5 HEMIACETALS AND HEMIKETALS

Aldehydes and ketones both contain a carbonyl oxygen (shown in bold in the margin). The carbon–oxygen bond in the carbonyl group is polar with the oxygen being electronegative. The carbonyl group is a site of chemical reactivity. An important reaction of aldehydes and ketones is with an alcohol, forming hemiacetals and hemiketals, as shown in Figure 7.7. Hemiacetals and hemiketals are unstable and the reaction is readily reversible at room temperature. Cyclic hemiacetals can result from a reaction between hydroxyl and carbonyl groups on the same chain. Those containing five or six atoms in a ring structure contrast markedly with other hemiacetals in that the formation of a ring is favoured. This is because the bond angles of the carbons in such rings is about 109° which is 'strain' free.

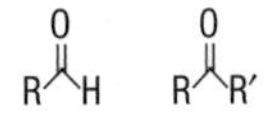

Aldehyde Ketone

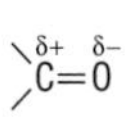

Polarity in carbonyl groups

The carbonyl carbon has been reduced to a hydroxyl in hemiacetal and hemiketal formation. Thus hemiacetals and hemiketals normally contain four different groups bound to one carbon (—H or —C; —C; —OH; and —O—C) and so are chiral centres. This new chiral centre is important when considering sugar cyclisation (see section 7.8).

Hemiacetals and hemiketals contain both a hydroxyl group and an ether linkage. The hydroxyl group can be further reacted with another alcohol to give acetals or ketals. The acetal or ketal linkage is often described as being a carbon containing two ether bonds. Acetal and ketal formation proceeds readily at room temperature because the hemiacetal hydroxyl is activated by the ether linkage. The resulting acetal or ketal is a stable structure which can easily be isolated in alkaline solution. The acetal linkages are less stable in acid solutions which allows a site for their breakdown.

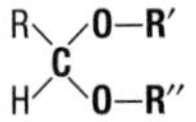

Acetal

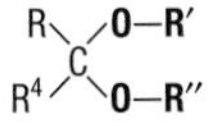

Ketal

Ether bonds shown in bold

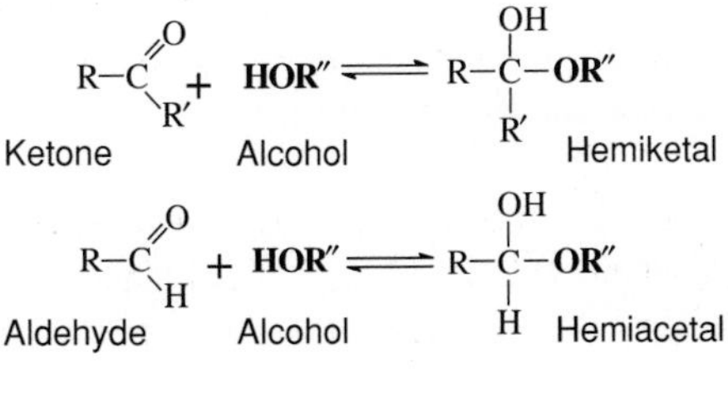

$$R{-}C(=O){-}R' + \mathbf{HOR''} \rightleftharpoons R{-}C(OH)(R'){-}\mathbf{OR''}$$

Ketone Alcohol Hemiketal

$$R{-}C(=O){-}H + \mathbf{HOR''} \rightleftharpoons R{-}C(OH)(H){-}\mathbf{OR''}$$

Aldehyde Alcohol Hemiacetal

• **Figure 7.7** Reaction of aldehydes or ketones with an alcohol forms hemiacetals and hemiketals.

Table 7.3 Naming sugars by reference to length of carbon chain and position of carbonyl carbon

Carbon chain	*Name*	*Carbonyl group*	
		Ketone	*Aldehyde*
3	Triose	Ketotriose	Aldotriose
4	Tetrose	Ketotetrose	Aldotetrose
5	Pentose	Ketopentose	Aldopentose
6	Hexose	Ketohexose	Aldohexose
7	Heptose	Ketoheptose	Aldoheptose

7.6 SIMPLE SUGARS

Sugars are often called simple carbohydrates. The word carbohydrate (hydrates of carbon) hints at a general formula of $(CH_2O)_n$ where n is 3 or more. The carbon atoms are normally in a linear chain and can be named by reference to the length of the carbon chain as shown in Table 7.3. It should be noted that sugars are identified by the suffix -ose. Each carbon on the chain contains one oxygen atom that is either in a carbonyl or a hydroxyl group. There is only **one** carbonyl group that can reside either at the end of the chain, making the group an aldehyde, or on one of the other carbons, making it a ketone or 'keto' sugar. Sugars could be more correctly termed polyhydroxyaldehydes or polyhydroxyketones. 'Aldo' sugars are called aldoses and an aldose containing six carbons is called an aldohexose, whilst 'keto' sugars are called ketoses and a ketose with five carbons is called a ketopentose.

WORKED EXAMPLE 7.5

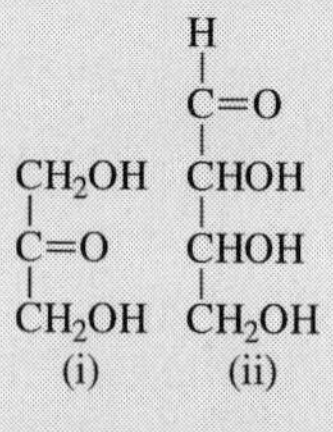

• **Figure 7.8** Two sugar molecules.

(i): CH_2OH–C=O–CH_2OH

(ii): H–C=O–CHOH–CHOH–CH_2OH

Identify whether the sugars shown in Figure 7.8 are aldoses or ketoses and name them by reference to the carbon chain length.

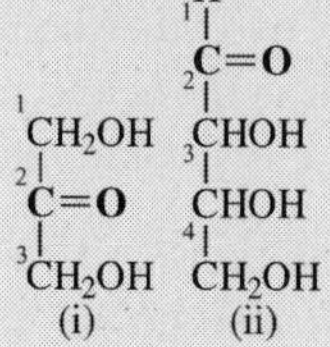

• **Figure 7.9** Two sugar molecules showing carbonyl oxygens in bold and carbon numbered.

(i): ${}^{1}CH_2OH$–^{2}**C=O**–${}^{3}CH_2OH$

(ii): H–^{1}**C=O**–^{2}CHOH–^{3}CHOH–${}^{4}CH_2OH$

ANSWER

Count the number of carbons (see Figure 7.9). Sugar (i) has three carbons and the sugar is a triose whilst sugar (ii) has four carbons and is a tetrose.

Identify the carbonyl oxygen. This is shown in bold in Figure 7.9. Molecule (i) is a ketose whilst molecule (ii) is an aldose.

Therefore (i) is a ketotriose and (ii) is an aldotetrose.

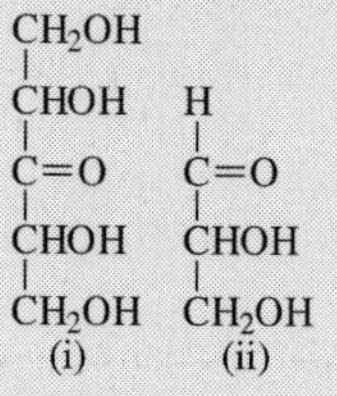

• **Figure 7.10** Two sugar molecules.

(i): CH_2OH–CHOH–C=O–CHOH–CH_2OH

(ii): H–C=O–CHOH–CH_2OH

QUESTION 7.5

Identify the sugars shown in Figure 7.10 by reference to carbonyl position and chain length.

The carbon atoms in sugars are numbered by reference to the position of the carbonyl group. In aldoses the aldehyde carbon is always carbon 1 whilst ketoses will be numbered by reference to the nearest end of the carbon chain where the carbonyl group is found. Thus the numbering of both the sugars in Figure 7.9 has been given. It is conventional to draw sugars in straight chain formation with carbon 1 at the top in Fisher projection.

WORKED EXAMPLE 7.6

Number the carbons in the sugars shown in Figure 7.11(a).

```
          H            CH2OH                CH2OH         H
          |              |                    |           |
         C=O            C=O                H-C-OH        C=O
          |              |                    |           |
       H-C-OH         H-C-OH        (b)      C=O       H-C-OH
          |              |                    |           |
(a)   HO-C-H         HO-C-H                H-C-OH     HO-C-H
          |              |                    |           |
       H-C-OH         H-C-OH              HO-C-H       H-C-OH
          |              |                    |           |
          H              H                  CH2OH       CH2OH
         (i)            (ii)                 (i)         (ii)
```

• **Figure 7.11** Some sugars. The carbonyl groups in (a) are shown in bold.

ANSWER

The carbonyl groups shown in Figure 7.11(a) are shown in bold. The carbons are simply numbered from the end of the carbon chain nearest to the carbonyl carbon. In both cases the carbon chains will be numbered with carbon 1 at the top.

QUESTION 7.6

Number the carbons in the sugars shown in Figure 7.11(b).

■ 7.7 CHIRALITY IN SIMPLE SUGARS

The figure in the margin shows the simplest aldo sugar glyceraldehyde (the aldehyde of glycerol). The central carbon, shown in bold, is bonded to four different groups. Thus it is a chiral centre and can exist in two forms. These two forms are left handed (2-hydroxyl to the left) known as L- or right handed (2-hydroxyl to the right) known as D-. The different chiral forms of a sugar are also called optical isomers or stereoisomers. In longer sugars the D- or L- is worked out by examining the furthest chiral carbon from the carbonyl group. The hydroxyl group attached to this carbon can be either to the right (D-) or to the left (L-). Carbons at the end of the chain (CH_2OH) and carbonyl carbons (C=O) are not joined to four different groups and are not chiral centres.

The number of stereoisomers of a sugar can be worked out by counting the number of chiral centres. For one chiral centre there are just two stereoisomers (left and right). When there are two chiral centres there are four possible stereoisomers (two at chiral centre 1 and two at chiral centre 2) as shown in the margin. In the figures in the margin the hydrogen and carbon atoms have been left out for clarity and the chiral centres have been indicated by an asterisk. The number of chiral forms can be calculated by using the equation:

$$\text{forms} = 2^n$$

where n is the number of chiral centres. The names of each of the chiral centres in both aldoses and ketoses can be found by reference to any good biochemistry textbook.

```
     H            H
     |            |
    C=O          C=O
     |            |
  H-C-OH      HO-C-H
     |            |
  H-C-OH       H-C-OH
     |            |
     H            H
    D-           L-
```

Isomers of glyceraldehyde

```
  =O            =O
 *-OH       HO-*
 *-OH          *-OH
  -OH           -OH

    =O          =O
HO-*           *-OH
HO-*        HO-*
   -OH          -OH
```

Four chiral forms of aldotetroses

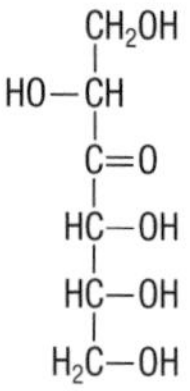

CH_2OH
HO—CH
C=O
HC—OH
HC—OH
H_2C—OH

WORKED EXAMPLE 7.7

The figure in the margin shows a ketohexose. (a) Indicate using an asterisk each chiral centre and (b) calculate how many stereoisomers exist for this ketohexose.

ANSWER

(a) Figure 7.12 shows the same ketohexose with the chiral centres indicated by an asterisk. The end carbons, shown in italics, are not chiral as each has bonded with two identical hydrogens. The carbonyl carbon, shown in bold, is not chiral as the carbon–oxygen is a double bond.

(b) There are three chiral centres. The number of isomers is 2^n where n is three for this sugar. 2^3 is 8 and so there are 8 isomers of this ketohexose.

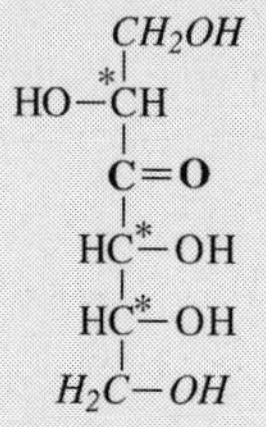

CH_2OH
HO—CH*
C=O
HC*—OH
HC*—OH
H_2C—OH

• **Figure 7.12**
A ketohexose with chiral centres marked with an asterisk.

QUESTION 7.7

The figure in the margin below figure 7.12 shows a ketopentose. (a) Indicate each chiral centre using an asterisk and (b) calculate how many stereoisomers exist for this ketopentose.

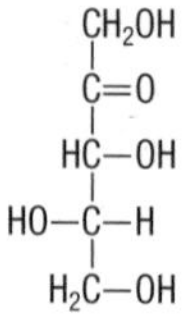

CH_2OH
C=O
HC—OH
HO—C—H
H_2C—OH

■ 7.8 STRAIGHT-CHAIN SUGARS SPONTANEOUSLY FORM RINGS

Sugars in which the carbonyl group and hydroxyl groups are four or five carbons apart are capable of bending so that a reaction takes place between the hydroxyl and carbonyl group. The product of such reactions in aldoses are hemiacetals and in ketoses are hemiketals (see section 7.5). The bond angles between carbons mean that only carbonyl and hydroxyl groups which are four or five carbons apart are capable of reacting in this way to form a stable intermediate. The cyclic structure resulting from a model compound of this type is shown in the margin. The reacting groups may be only four carbons apart, in which case a five-membered ring containing one oxygen, known as a furan ring, is produced. Alternatively, the reacting groups may be five carbons apart, in which case a six-membered ring containing an oxygen (called a pyran ring) is produced. The pyran ring is shown in the margin. The hemiacetal or hemiketal bond produced is easily broken and is in equilibrium with the open chain form.

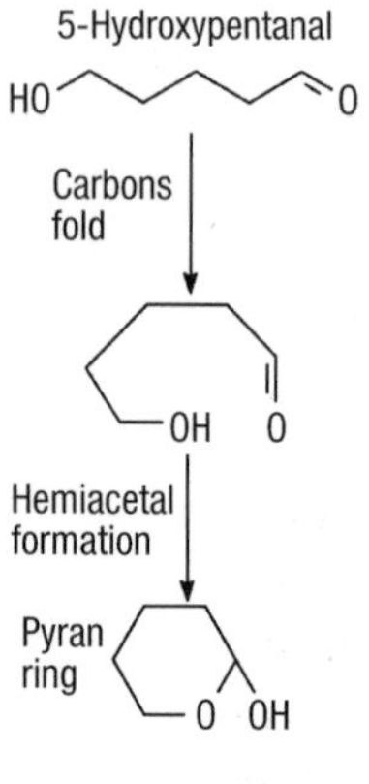

5-Hydroxypentanal
HO O
Carbons fold
OH O
Hemiacetal formation
Pyran ring
O OH

Sugars containing furan or pyran rings are known as furanoses or pyranoses respectively. Thus glucose can cyclise as either a furanose or pyranose structure as given in Figure 7.13. The carbonyl group is reduced to a hydroxyl group by this reaction and the carbon to which it is attached becomes a chiral centre. The chiral carbon formed as a result of ring formation is called the anomeric carbon. In the Haworth projection of the formula shown in Figure 7.13 the hydroxyl derived from the carbonyl oxygen is trapped in the 'up' position and the result is the β-anomer. The hydroxyl can also be trapped in the 'down' position to produce the α-anomer. The difference between α- and β-anomers is shown in the margin. The difference between these anomers may not seem great but when sugars are joined together the consequences can be marked.

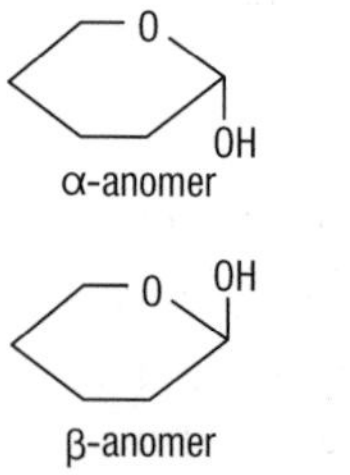

O
OH
α-anomer

O OH
β-anomer

Anomeric carbons contain hydroxyl groups in opposite configuration

The numbering of the carbons in the straight chain formula is kept when sugars cyclise.

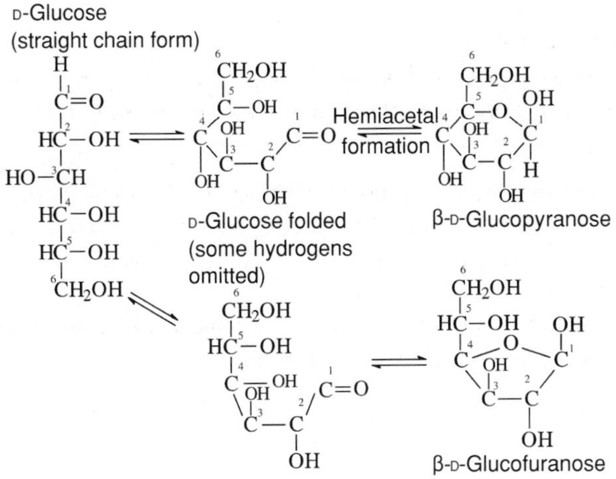

D-Glucose (straight chain form)

D-Glucose folded (some hydrogens omitted)

Hemiacetal formation

β-D-Glucopyranose

β-D-Glucofuranose

• **Figure 7.13** Formation of furanose and pyranose rings from β-D-glucose.

Thus carbon 1 in aldoses remains carbon 1 whether the aldose forms a furan or pyran ring. The numbering system is shown in Figure 7.13 for glucose.

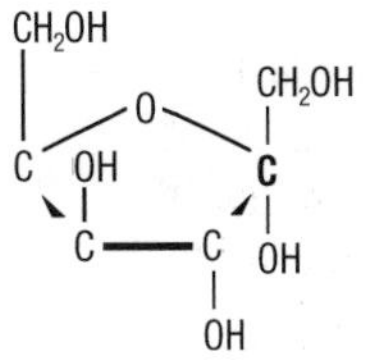

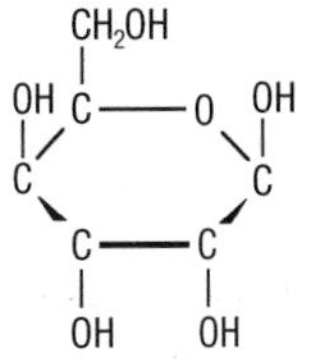

WORKED EXAMPLE 7.8

Number the carbons in the sugar shown in the margin.

ANSWER

Identify the anomeric carbon. This is the carbon next to the ring oxygen which also has a hydroxyl group attached. This is given in bold in the marginal figure. Then number the carbon chain (in both directions!). The numbering which gives the anomeric carbon the lowest number is correct. In the example shown the anomeric carbon is carbon 2.

QUESTION 7.8

Number the carbons in the sugar shown in the margin.

■ 7.9 SUGAR HYDROXYLS CAN BE CHEMICALLY MODIFIED

The hydroxyl groups in simple sugars are capable of undergoing chemical modification to form important sugar derivatives. Figure 7.14 shows some of the types of chemical modification that the glucose molecule can undergo and some of the derivatives that can be formed. Other important derivatives include sugar alcohols. Sugar alcohols have the carbonyl oxygen reduced to a hydroxyl group. One of the most widely used is **sorbitol** which is well metabolised by the human body and is therefore often used in diabetic products. It is widely used as a sweetener by the food industry as it has superior food processing properties to sucrose, but has a laxative effect that limits its use. Sugar phosphates are also very important and are covered in Chapter 9.

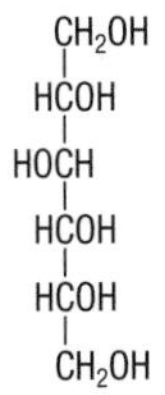

CH_2OH
HCOH
HOCH
HCOH
HCOH
CH_2OH

D-Sorbitol

■ 7.10 SUGARS ARE JOINED TOGETHER BY GLYCOSIDIC BONDS

The hydroxyl group from one sugar molecule may be covalently linked to the hydroxyl group from another sugar to form an **ether linkage** (see margin). The ether bond between sugars is called a glycosidic bond. Sometimes the anomeric hydroxyl is involved in glycosidic bond formation. In this case the group formed is an acetal or ketal. Acetals and ketals are much more stable than hemiacetals and hemiketals and so ring sugars do not spontaneously

two methanols

$$CH_3OH + HOCH_3 \rightleftharpoons CH_3OCH_3 + H{-}O{-}H$$

Ether Water

Formation of an ether linkage between two hydroxyls

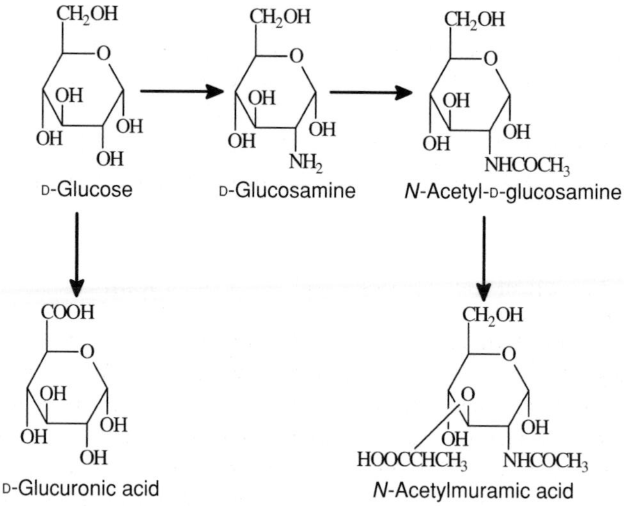

• **Figure 7.14** Some reactions that derivatise glucose (some hydrogens left out for clarity).

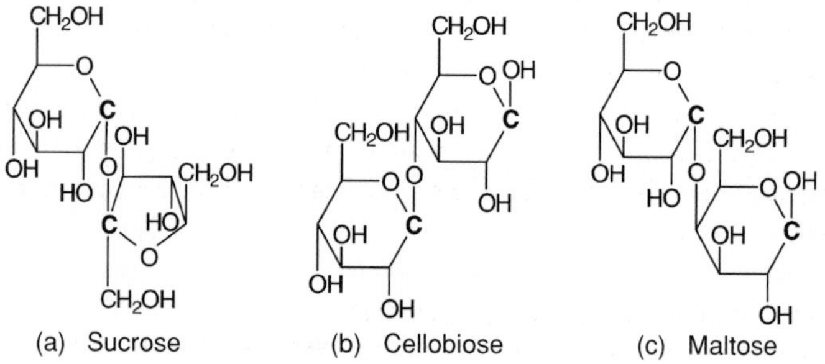

D-Glucose → D-Glucosamine → *N*-Acetyl-D-glucosamine

D-Glucuronic acid

N-Acetylmuramic acid

• **Figure 7.15** Structures of some disaccharides showing anomeric carbons in bold.

(a) Sucrose (b) Cellobiose (c) Maltose

open up after acetal or ketal formation. The sugar carbonyl group cannot then be oxidised by molecules such as Cu^{2+} (see Chapter 13) after acetal or ketal production. The reduction of Cu^{2+} ions by sugars forms the basis of both the Fehlings and Benedicts test for sugars. Such sugars are said to be non-reducing after glycosidic bonds have been formed. In practice most sugars do not link via both anomeric hydroxyls and so many polymers of sugars retain at least one end capable of acting to reduce ions such as Cu^{2+}. A notable exception to this is sucrose. The structure of sucrose is shown in Figure 7.15(a) with the anomeric carbons shown in bold. The structure of dimers of glucose, maltose and cellobiose, are shown in Figure 7.15(b) and (c) with the anomeric carbons shown in bold.

Polymers of sugars are called saccharides. Sucrose, maltose and cellobiose are disaccharides as two sugar monomers are linked together. Three, four or many sugars linked together are called tri-, tetra- or polysaccharides, respectively.

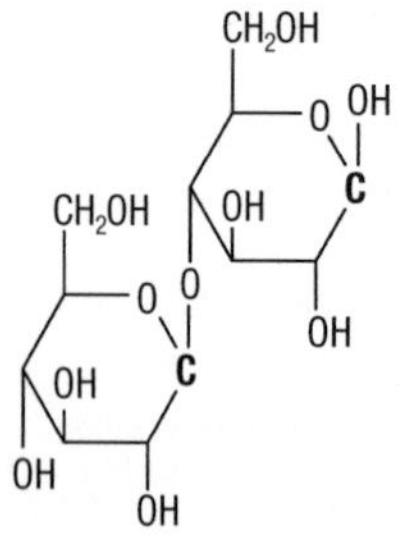

WORKED EXAMPLE 7.9

Decide whether the disaccharide shown in the margin is reducing or non-reducing.

ANSWER

Identify the anomeric carbons. Identification of the anomeric carbon is easy if you look for the one of the two carbons next to the oxygen in the ring that has another oxygen directly bound to it. These are shown in bold for this sugar.

The sugar is reducing if one of the anomeric carbons is at the end of the sugar. This sugar has one end that is capable of acting as a reducing sugar.

QUESTION 7.9

Decide whether the disaccharide shown in the margin is reducing or non-reducing.

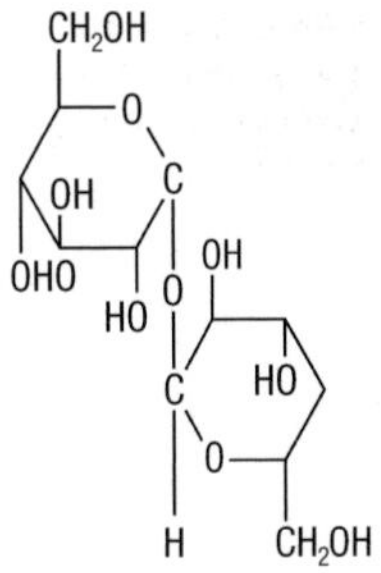

■ SUMMARY

Fatty acids are long-chain carboxylic acids. The alkyl group can be saturated or contain alkene groups in the *cis* conformation. The melting temperature of fatty acids is affected by chain length and degree of unsaturation. The fatty acids can be esterified with glycerol to form triglycerides, the main component of many fats and oils. Reactions between aldehydes or ketones and alcohol groups yield reactive hemiacetals or hemiketals from which stable acetals and ketals can be formed. Sugars are polyhydroxyaldehydes or polyhydroxyketones. In sugars ring structures can form as a result of hemiacetal or hemiketal formation. Two cyclic isomers only are produced; a five-membered ring containing one oxygen called a furanose and a six-membered ring containing one oxygen known as a pyranose. The cyclisation process introduces a new chiral centre into the sugar, known as the anomeric carbon. The other hydroxyls are capable of modification to yield sugar derivatives such as sugar amines. Sugars can be linked together by ether bonds to form saccharides.

■ SUGGESTED FURTHER READING

Lehninger, A.L., Nelson, D.L. and Cox, M.M. (1993) *Principles of Biochemistry*, pp. 240–267, 298–323, 2nd edn. Worth, New York.

Excellent detailed coverage of the range of fats, oils, lipids, sugars and saccharides is also given in most undergraduate biochemistry textbooks.

Gurr, M.I. and Harwood, J.L. (1991) *Lipid Biochemistry*, 4th edn. Kluwer, Dordrecht.

Candy, D.J. (1980) *Biological Functions of Carbohydrates*. Kluwer, Dordrecht.

■ END OF CHAPTER QUESTIONS

Question 7.10 Name the unsaturated fatty acid shown in Figure 7.16.

• Figure 7.16 The structure of a fatty acid.

Question 7.11 Draw and name the ester formed between ethanoic acid and methanol.

Question 7.12 (a) Explain what is meant by the term anomeric carbon.

(b) Identify the sugars shown in Figure 7.17 as α- or β-anomers.

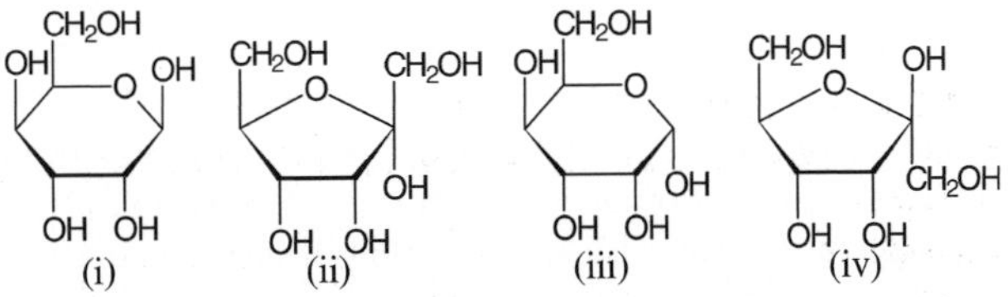

(i) (ii) (iii) (iv)

• Figure 7.17 Some sugars.

Question 7.13 Sucrose contains an acetal and a ketal group.

(a) Draw the structure of sucrose and label the acetal and ketal groups.

(b) Explain why sucrose is not a reducing sugar.

8

ORGANIC AND BIOLOGICAL REACTION MECHANISMS

■ 8.1 INTRODUCTION

Metabolic pathways within organisms require a sequence of consecutive reactions to take place. These are mediated by enzymes and involve an apparently complex range of changes to biomolecules. The processes can be understood in terms of the reactions of simple organic compounds. Substitution, addition and elimination may take place at specific reactive centres within a molecule. An understanding of the charge distribution around a chemical bond and a recognition of such centres allows simple reactions to be interpreted. The reactions of large biomolecules can then be explained by analogy with model organic molecules.

■ 8.2 REACTIVE SITES AND FUNCTIONAL GROUPS

Functional groups with polar bonds form reactive sites

When a relatively small molecule, for example butanol, $CH_3CH_2CH_2CH_2OH$, is considered, there are a large number of atoms or bonds at which a chemical reaction could take place. Fortunately, experience has shown that the **reactive site** in a molecule is almost always the functional group. In this case, it is the hydroxyl group, —OH, that provides the centre for reactivity; it is the reactive site. The functional group behaves in this way because it is different from the rest of the molecule in having polar bonds. In butanol, the hydroxyl group is polarised (section 3.3) as follows:

$$\overset{\delta+}{C}-\overset{\delta-}{O}H$$

A lone pair of electrons is slightly negative and acts as a nucleophile

A nucleophile is attracted to, and reacts with, a positive atom

An electrophile is slightly positive; it is attracted to, and reacts with, a negative centre

This means that the oxygen atom will attract reagents with a positive charge or a slightly positive centre. In the same way, the carbon atom will be attracted by reagents with a negative charge or a slightly negative centre. By contrast, the carbon–carbon and carbon–hydrogen bonds in the remainder of the molecule are non-polar or only very slightly polar and do not attract charged reagents. Another significant feature of the hydroxyl group is the lone pair of electrons on oxygen (section 2.9). A lone pair of electrons is like a finger of negative charge protruding from the molecule. The oxygen carries two lone pairs of electrons, each of which is strongly attracted to a positive site. An oxygen of this type is termed a **nucleophilic centre**. It is possible to identify nucleophilic centres in a number of functional groups. In the same way the slightly positive carbon atom bound to oxygen is known as an **electrophilic centre**.

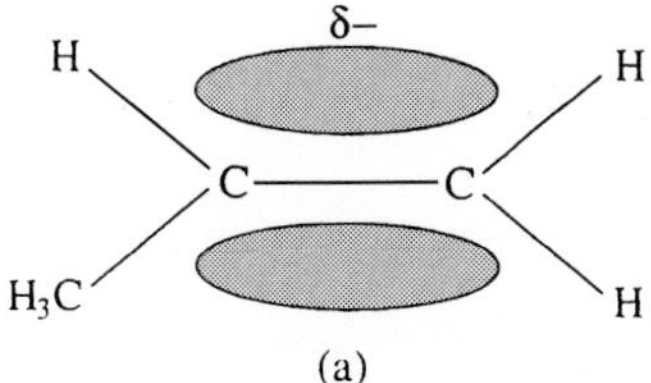

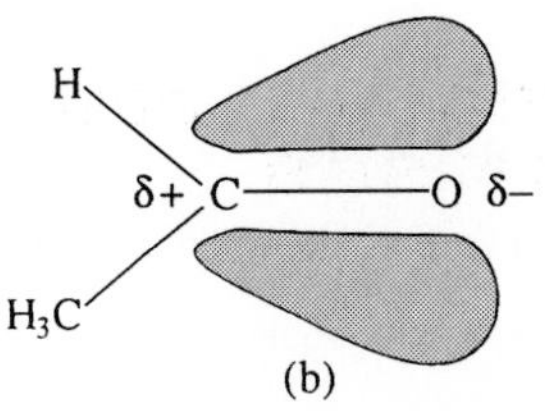

• **Figure 8.1** (a) Non-polar and (b) polar π bonds in propene and ethanal.

A number of functional groups contain double bonds. The two separate bonds making up the double bond are of different types. The first bond consists of a pair of electrons held tightly between the two atoms – it is a σ bond (sigma bond) (section 2.6). The second bond is formed by a pair of electrons located in charge clouds above and below the internuclear axis of the bond. It is a π bond (pi bond) (section 2.7). The π electrons are less tightly held than the σ electrons and so can behave as reactive sites and attract other reactive centres. Double bonds can be non-polar, as in alkenes such as propene, $CH_3CH{=}CH_2$, or polar, as in ethanal, CH_3CHO. The non-polar double bond in propene carries a small negative charge on the π-electron cloud (Figure 8.1(a)) and so it is attracted to electrophilic or positive centres. The carbon–oxygen double bond in ethanal is polarised with a slightly positive carbon atom and a slightly negative oxygen (Figure 8.1(b)). Nucleophiles are attracted to the carbon atom.

The **Lewis acid–base theory** is often used to rationalise the idea of reactive sites. A **Lewis acid** is regarded as a substance that will accept an electron pair. By contrast, a **Lewis base** is a substance that will donate an electron pair. In this way, the carbon atom bound to oxygen in butanol is a Lewis acid site, as is the carbon in the carbon–oxygen double bond of ethanal. The π electrons in the double bond of propene provide a Lewis base site as do the lone pair electrons on the oxygen atom in butanol. Before listing some important reactive sites in organic molecules it is useful to summarise the concepts discussed so far.

A Lewis acid will accept electrons

A Lewis base will donate electrons

There are three types of electron pairs in organic and biomolecules that can take part in chemical reactions.

- σ-bond pairs — these electrons are tightly held in the bond and show low reactivity, they have low Lewis basicity,
- π-bond pairs — π electrons show more delocalisation, they have medium reactivity and medium Lewis basicity,
- lone pairs — the electrons protrude into space, they have high reactivity and high Lewis basicity.

A list of reactive and unreactive sites within organic molecules is collected in Table 8.1.

WORKED EXAMPLE 8.1

(a) Draw the structural formula of 1-aminoprop-2-ene.

(b) Identify:

 (i) a Lewis acid centre,

 (ii) a Lewis base centre; and indicate the charge on it.

(c) Mark the position of π electrons.

(d) Show the position of lone pair electrons.

(e) Indicate an unreactive σ bond.

ANSWER

(a) $CH_2{=}CH{-}CH_2{-}NH_2$

(b) (i) The carbon atom bound to nitrogen is a Lewis acid centre $CH_2{=}CH{-}\overset{\delta+}{C}H_2{-}NH_2$

(ii) The electronegative nitrogen atom is a Lewis base $CH_2{=}CH{-}CH_2{-}\overset{\delta-}{N}H_2$

(c) The carbon–carbon double bond carries π electrons $CH_2\overset{\pi}{=}CH{-}CH_2{-}NH_2$

(d) The nitrogen atom has a lone pair of electrons $CH_2{=}CH{-}CH_2{-}\ddot{N}H_2$

(e) The carbon–carbon single bond or any one of the carbon–hydrogen bonds are unreactive σ bonds

It should be noted that the π electrons identified in (c) form a region of high electron density and so may be regarded as a Lewis base centre. The lone pair electrons in (d) have a partial negative charge and so form the Lewis base site marked in (b)(ii).

Table 8.1 Reactive and unreactive sites in organic molecules

Type of site	*Name*	*Formula*
Reactive, Lewis acid, slightly positive carbon atom	Alcohol	$\overset{\delta+}{C}{-}\overset{\delta-}{OH}$
Reactive, Lewis acid, slightly positive carbon atom	Amine	$\overset{\delta+}{C}{-}\overset{\delta-}{NH_2}$
Reactive, Lewis acid, slightly positive carbon atom	Alkyl halide	$\overset{\delta+}{C}{-}\overset{\delta-}{Cl}$
Reactive, Lewis acid, slightly positive carbon atom	Aldehyde, ketone, carboxylic acid, ester	$\delta{+}C{=}O\ \delta{-}$
Reactive, Lewis base, slightly negative, lone pair electrons	Alcohol	$\overset{\delta+}{C}{-}\overset{\delta-}{O}$
Reactive, Lewis base, slightly negative, lone pair electrons	Amine	$\overset{\delta+}{C}{-}\overset{\delta-}{N}$
Reactive, Lewis base, slightly negative, π electrons	Alkene	$\overset{\delta-}{C{=}C}$
Reactive, Lewis base, slightly negative, π electrons and lone pair electrons	Aldehyde, ketone, carboxylic acid, ester	$\delta{+}C{=}O\ \delta{-}$
Unreactive, non-polar single bond, σ electrons	Alkane, alkyl group	$C{-}C$
Unreactive, slightly polar single bond, σ electrons	Alkane, alkyl group	$C{-}H$

■ 8.3 DESCRIBING REACTION MECHANISMS

A number of terms and conventions are used in order to discuss reactions and reaction mechanisms.

Nucleophile This is a molecule or anion with a lone pair of electrons which acts as a Lewis base. It is often written:

Nu⟨⇅⟩ or alternatively Nu:

A reaction involving a nucleophile is called a **nucleophilic reaction**.

Electrophile This term is used to describe a molecule or cation with an electron deficient site (+ or δ+) which acts as a Lewis acid. It can be written: E. A reaction involving an electrophile is an **electrophilic reaction**.

When reactions involve either nucleophiles or electrophiles then the electrons involved in forming new bonds, and in breaking existing bonds, move in pairs.

Free radical When an organic species carries an unpaired electron it is called a free radical. Such a radical is uncharged and does not specifically seek out positive or negative reactants. Reactions involving free radicals are termed **free radical reactions**.

A free radical has an atom with an unpaired electron

A free radical is often written as a formula followed by a superscript dot. In this way, the methyl free radical is written: $CH_3^{\cdot}$

A simple reaction mechanism is described by:

- drawing the reactant molecule or ion showing the reactive sites in the correct orientations with respect to one another;
- adding a curly arrow, ↷, or arrows, to show the movement of a pair, or pairs, of electrons in the case of nucleophilic or electrophilic reactions; a half arrow, ↷, or arrows, is added in the case of free radical reactions to indicate the movement of a single electron, or electrons;
- drawing the activated complex, or transition state, in square brackets with any necessary charges;
- drawing the product species.

WORKED EXAMPLE 8.2

Identify from the following list of reagents:

(a) an electrophile,

(b) a nucleophile,

(c) a free radical.

List of reagents:

(i) the cyanide ion, CN^- (ii) the proton, H_3O^+

(iii) ethane, CH_3CH_3 (iv) a chlorine atom, $Cl^{\cdot}$

ANSWER

(a) The electrophile is the positive Lewis acid species (ii).

(b) The nucleophile is the anionic cyanide ion (i).

(c) The free radical has an unpaired electron and is (iv).

QUESTION 8.1

Identify:

(a) a nucleophilic site in the alcohol methanol, CH_3OH,

(b) a free radical site in the species $CH_3CHCH_2CH_3$,

(c) an electrophilic site in the aldehyde ethanal, CH_3CHO.

8.4 BIMOLECULAR NUCLEOPHILIC SUBSTITUTION

Bimolecular nucleophilic substitution allows one group to replace another in a molecule

This reaction occurs when a nucleophile with a lone pair of electrons attacks a saturated carbon atom which is electron deficient. The nucleophile becomes attached to the electron deficient carbon, while a group initially bound to it is displaced. One group substitutes for another. The mechanism is called bimolecular because two molecules take part in the transition state. An example from organic chemistry will illustrate how the mechanism is represented. Chloroethane, CH_3CH_2Cl, reacts with the hydroxide ion in sodium hydroxide, when heated in water, to give ethanol, CH_3CH_2OH, and sodium chloride. The mechanism is represented in Figure 8.2. It should be noted that in the figure a lone pair of electrons is transferred to an electrophilic carbon atom while the electrons in the carbon–chlorine bond are transferred to the chlorine atom. The reactants are in equilibrium with the transition state (ts) enclosed by square brackets. The bonds that are breaking or forming are shown dotted in the transition state. The carbon atom at the centre of the transition state has three atoms, or groups, linked to it by normal two-electron bonds. A further two groups are linked by partial bonds and so the carbon has a total of five groups around it. Thus the transition state can exist for only a very short time. It must either rapidly form the products or go back to the reactants. The configuration around the carbon atom, at which reaction occurs, undergoes inversion during the reaction.

Nucleophilic substitution in biomolecules is rather different. In place of the powerful chemical reagent (sodium hydroxide) and forcing conditions (heating to boiling point), a subtle enzyme catalyst allows the reaction to occur at a pH near to 7 and at around room temperature. The group to be displaced is not usually the easy-to-displace chloride leaving group but a less easily substituted unit such as alkoxide. The alkoxide group is made up of an alkyl group linked to an oxygen atom. Thus the glycosidic linkage in the storage polysaccharide glycogen can be hydrolysed with inversion of configuration at the reactive centre using the enzyme glycosidase (Figure 8.3). The reaction is more complex than

• **Figure 8.2** Bimolecular nucleophilic displacement of chloride from chloroethane by the hydroxide ion.

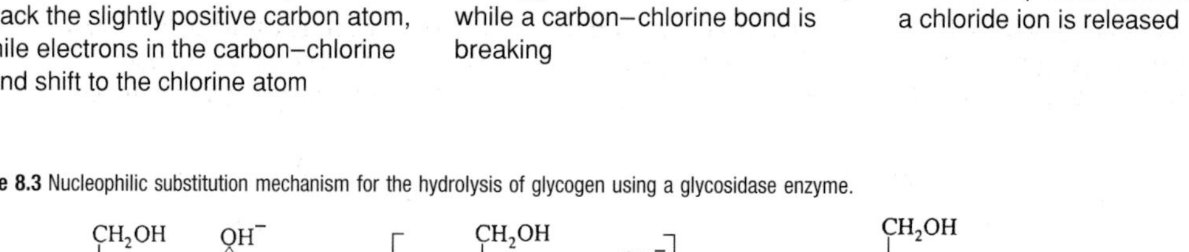

Lone pair electrons on the hydroxide ion attack the slightly positive carbon atom, while electrons in the carbon–chlorine bond shift to the chlorine atom

An oxygen–carbon bond is forming while a carbon–chlorine bond is breaking

An alcohol product is formed and a chloride ion is released

• **Figure 8.3** Nucleophilic substitution mechanism for the hydrolysis of glycogen using a glycosidase enzyme.

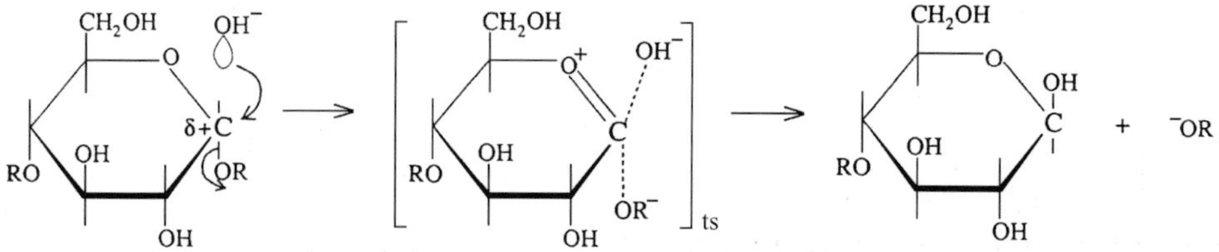

β-Glycoside

The reactive carbon has three full bonds and two partial bonds

α-Glycoside

Alkoxide ion

the straightforward nucleophilic displacement described for chloroethane. Hydrolysis of glycogen can also take place without inversion; a β-glycosidase enzyme breaks the β-glycosidic linkage and an α-glycosidase will cleave an α-linkage. Inversion of configuration occurs when one mirror image form of a molecule is converted to the opposite mirror image form.

QUESTION 8.2
Explain briefly the meaning of the terms:
(a) nucleophilic substitution,
(b) transition state.

■ 8.5 ELECTROPHILIC ADDITION TO A NON-POLAR DOUBLE BOND

The reaction requires an electron-deficient centre, an electrophile, to attack a π-electron pair, a Lewis base centre. This is usually a carbon–carbon double bond in an alkene or alkene derivative. The mechanism takes place in two stages. Initially, the electrophile receives a pair of electrons from the π bond to form a positive transition state in which two electrons are shared between three atoms. In the second stage, the transition state is attacked by a nucleophile derived from the initial electrophilic reagent. The process can be exemplified by the addition of hydrogen bromide to ethene which forms bromoethane (Figure 8.4). Hydrogen bromide is a more powerful electrophile than will normally be encountered in the biosphere. Electrophilic addition of the mild reagent water to an alkene double bond is significant in more than one stage of cellular respiration. Citrate is isomerised

Electrophilic addition allows groups to add to each end of an alkene double bond

• **Figure 8.4** Electrophilic addition of hydrogen bromide to ethene forming bromoethane.

π electrons in the double bond are attracted to the slightly positive hydrogen atom. Electrons in the hydrogen–bromine bond shift on to the bromine

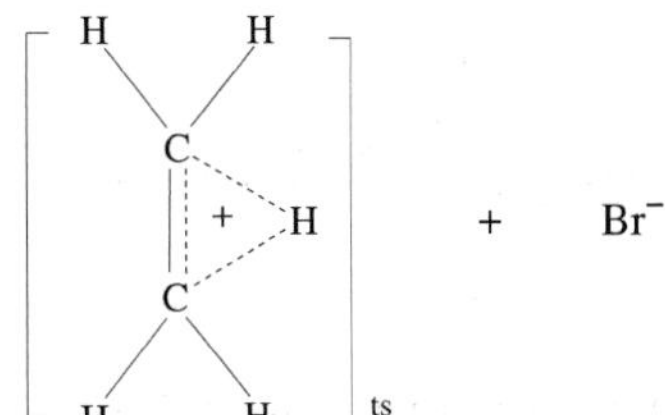

The two carbon atoms each have three other atoms linked by normal bonds and two atoms joined by partial bonds

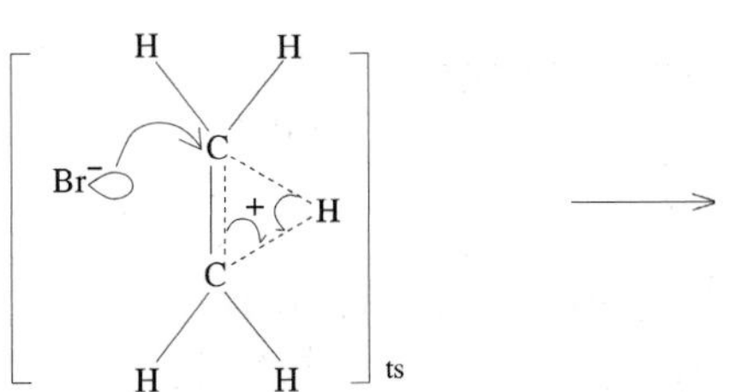

Lone pair electrons on bromine attack a slightly positive carbon atom. Two one-electron shifts form a new carbon–hydrogen bond

A bromoalkane product is formed

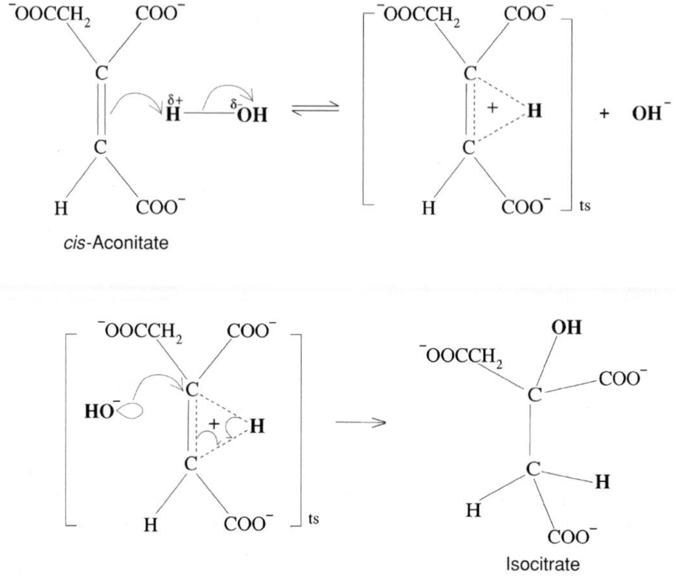

• **Figure 8.5** Electrophilic addition of water to *cis*-aconitate catalysed by the enzyme aconitase to form isocitrate.

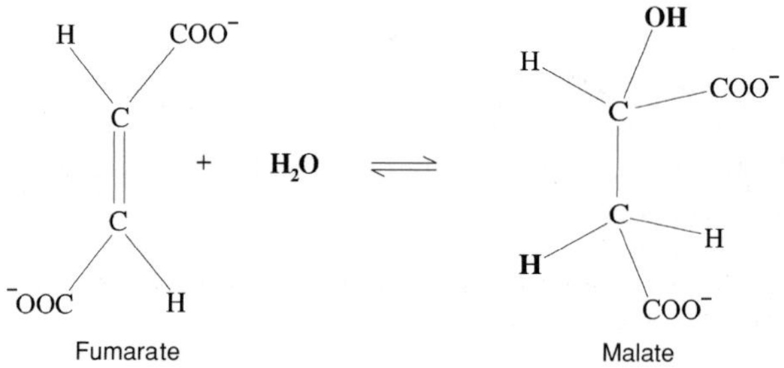

cis-Aconitate

Isocitrate

• **Figure 8.6** Electrophilic addition of water to fumarate to give malate. The reaction is catalysed by the enzyme fumarase.

Fumarate

Malate

to isocitrate by way of the intermediate *cis*-aconitate. This intermediate is an alkene which undergoes electrophilic addition of water to form isocitrate under the influence of the enzyme aconitase (Figure 8.5). In a later stage of the citric acid cycle, fumarate undergoes conversion to malate by electrophilic addition of water to the double bond; the enzyme involved is fumarase (Figure 8.6).

QUESTION 8.3

Name two enzymes involved in cellular respiration that use electrophilic addition in the reaction mechanism. Use a reaction scheme to show the mechanism of one of these reactions.

8.6 ELIMINATION TO FORM AN ALKENE

In elimination, a group is lost from each of two neighbouring carbon atoms to form an alkene double bond

Addition of water to a double bond is often closely linked to elimination of water in metabolic reaction sequences. Elimination takes place mainly by one of two mechanisms in biological systems. **Concerted elimination** occurs in a single-stage process by which a base removes a proton from one side of the molecule while a hydroxide ion leaves

• **Figure 8.7** Mechanisms of elimination.

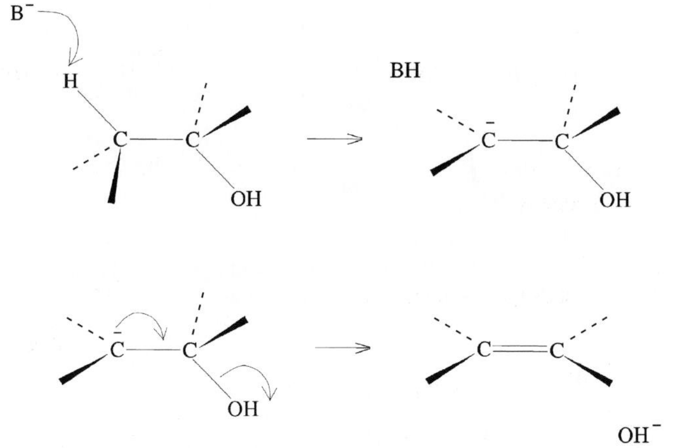

(a) Two-stage elimination of water to form an alkene. Proton abstraction occurs to form an intermediate anion which subsequently loses a hydroxide ion.

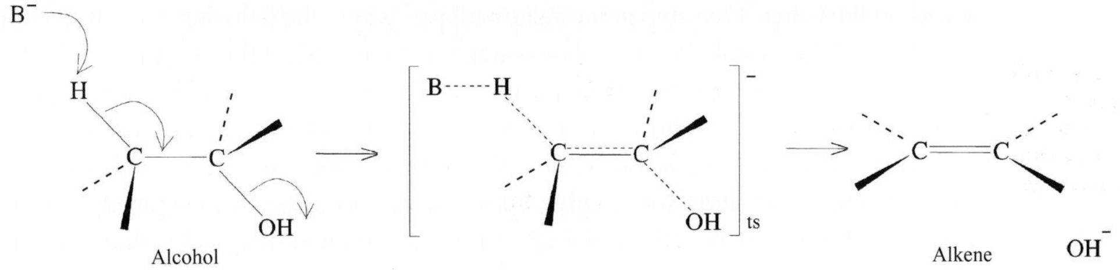

(b) Concerted elimination of water. A proton is abstracted while at the same time a hydroxide ion is lost.

from the other side. This means that the stereochemical course of the reaction is controlled to give an alkene of specific geometry (Figure 8.7). The alternative mechanism, **carbon–hydrogen cleavage**, takes place in two stages. The first step requires cleavage of a carbon–hydrogen bond followed in a separate step by loss of hydroxide (Figure 8.7). By this process the stereochemistry of the product is not fixed but depends on the nature of the active site in the catalytic enzyme involved. An example of concerted elimination in organic chemistry is the formation of but-2-ene from 2-bromobutane using sodium hydroxide in hot aqueous ethanol (Figure 8.8). A hydroxide ion acts as a base in abstracting a proton from a carbon–hydrogen group by transferring a pair of electrons to it. At the same time, the bromine atom on the opposite side of the molecule receives a pair of electrons from the carbon–bromine bond and leaves as a bromide ion. The product is the *cis*- or *trans*-isomer of butene. Enzyme-catalysed dehydration in biochemical pathways often follows a concerted mechanism when β-hydroxycarboxylic acids, also called 3-hydroxycarboxylic acids, $RCH(OH)CH_2COOH$, are the substrates. Two-stage elimination usually, but not invariably, occurs in the reactions of β-hydroxyketones and β-hydroxythioesters which are also called 3-hydroxyketones, $RCH(OH)CH_2COR'$, and 3-hydroxythioesters, $RCH(OH)CH_2COSR'$.

• **Figure 8.8** Concerted elimination of hydrogen bromide from a bromoalkene to form an alkene.

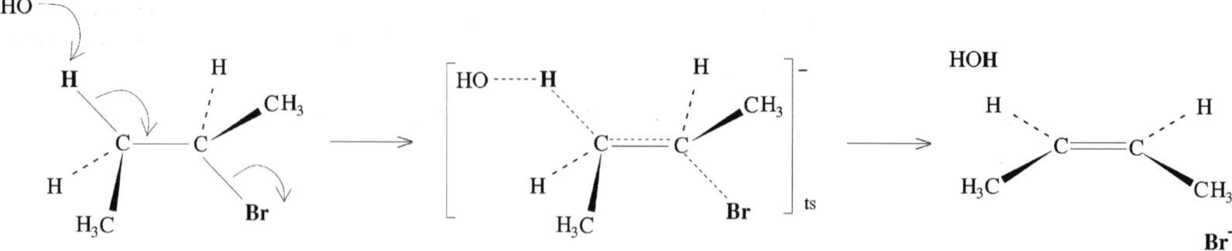

Lone pair electrons on the hydroxide ion move towards a hydrogen atom. Electrons in the carbon–hydrogen bond shift to the carbon–carbon bond. The bromine atom attracts electrons from the adjacent bond

Hydroxide–hydrogen and carbon–carbon bonds are forming while carbon–hydrogen and carbon–bromine bonds are breaking

An alkene is formed as water and bromide ion are liberated

The biosynthesis of aromatic aminoacids such as L-phenylalanine occurs in plants by a series of elimination reactions which are grouped together as the shikimate pathway. The route is significant since it has no parallel in animals, which have to seek these amino acids in their diet. One step in the shikimate pathway is the dehydration of 3-dehydroquinic acid to 3-dehydroshikimic acid. The reaction is mediated by the enzyme 3-dehydroquinate dehydratase. Although the mechanism is not a simple one, detailed investigations have shown that two-stage elimination occurs to give the alkene product (Figure 8.9).

This two-stage mechanism is involved in the reversible dehydration of β-hydroxydecanoyl thioesters to give *trans*-2-decenoyl thioesters (Figure 8.10). The reaction is catalysed by β-hydroxydecanoyl thioester dehydratase and is one of the reactions involved in the fatty acid synthase cycle.

Although ammonia is not a good leaving group in an elimination reaction, enzyme catalysis makes the process possible. Several L-amino acids, such as phenylalanine, undergo this process by using an ammonia lyase enzyme. The elimination takes places by way of a concerted mechanism to give a cinnamate product (Figure 8.11).

• **Figure 8.9** Elimination of water from 3-dehydroquinate to form 3-dehydroshikimate using the enzyme 3-dehydroquinate dehydratase. The mechanism is a two-stage process.

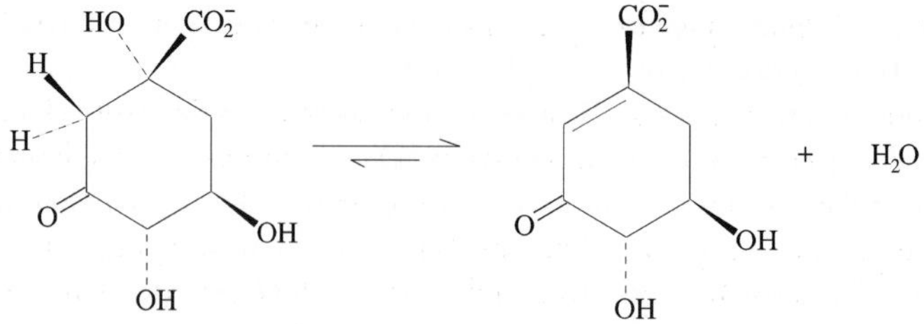

HO CO_2^- H H O OH OH ⇌ CO_2^- O OH OH + H_2O

3-Dehydroquinate 3-Dehydroshikimate

• **Figure 8.10** Enzyme-catalysed elimination of water from β-hydroxydecanoyl thioesters to give *trans*-2-decenoyl thioester. A two-stage mechanism is involved.

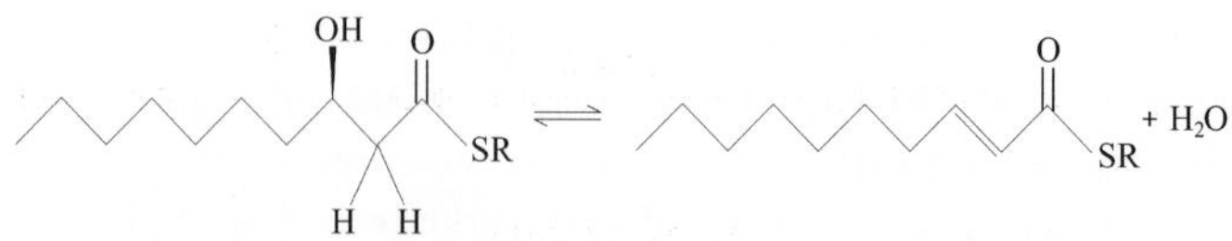

OH O SR H H ⇌ O SR + H_2O

β-Hydroxydecanoyl thioester *trans*-2-Decenoyl thioester

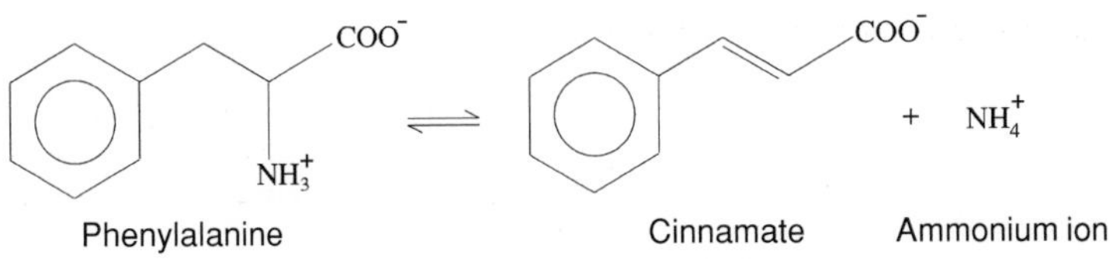

Phenylalanine ⇌ Cinnamate + Ammonium ion

COO^-, NH_3^+ ⇌ COO^- + NH_4^+

• **Figure 8.11** Concerted elimination of ammonia from phenylalanine catalysed by phenylalanine ammonia lyase.

QUESTION 8.4

(a) Explain carefully the difference between 'two-stage elimination' and 'concerted elimination'.

(b) Give one example of each type of elimination.

8.7 NUCLEOPHILIC ADDITION TO A POLAR DOUBLE BOND

A polar, carbonyl double bond can undergo addition followed by elimination

Hydrolysis reactions are of major significance in cellular metabolism. Hydrolytic enzymes break down biological food materials into smaller molecules which are then absorbed and used for the production of energy or in biosynthesis. Hydrolysis often involves attack by a nucleophile at the electropositive (Lewis acid) end of a polar double bond, frequently a carbonyl group. The addition product formed may then undergo elimination of a group originally bound to the electropositive carbon to reform the double bond. Closely related to hydrolysis is the group transfer reaction by which a group is removed from a substrate and transferred to an acceptor other than water. Hydrolysis and group transfer are compared in Figure 8.12.

Organic chemistry provides a well-known model for the nucleophilic addition to a double bond in the hydrolysis of an ester. The reaction uses an alkali, like sodium hydroxide, which is heated with the ester in water to give, as products, an alcohol and a carboxylic acid in the form of its sodium salt. The mechanism requires the hydroxide nucleophile to use a lone pair of electrons to attack the carbon atom of the carbonyl–oxygen double bond. This forms an intermediate anion-addition species which then undergoes elimination to reform the double bond and release the alcohol and carboxylic acid salt as the products. The process is summarised in Figure 8.13 for the ester ethyl ethanoate which is hydrolysed by aqueous sodium hydroxide to give ethanol and sodium ethanoate.

The peptide bond adjacent to an aromatic amino acid is selectively cleaved by water using the enzyme chymotrypsin, a serine protease. The enzyme catalysis involves the intercession of three active residues working together to bring about the hydrolysis. A simplified outline of a key stage of the mechanism is given in Figure 8.14. An enzyme-linked hydroxyl attacks the carbonyl double bond at the electropositive carbon atom to form an intermediate anionic addition species. This anion then undergoes elimination to form a free carboxylic acid and an amine and regenerate an enzyme-bound hydroxyl group.

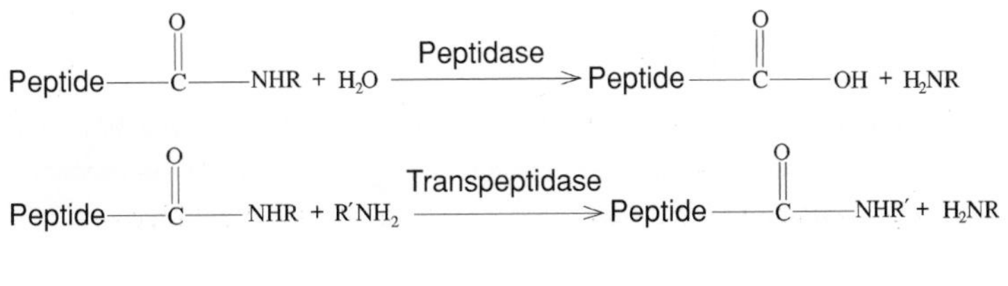

$$\text{Peptide}-\overset{O}{\overset{\|}{C}}-\text{NHR} + H_2O \xrightarrow{\text{Peptidase}} \text{Peptide}-\overset{O}{\overset{\|}{C}}-\text{OH} + H_2NR$$

$$\text{Peptide}-\overset{O}{\overset{\|}{C}}-\text{NHR} + R'NH_2 \xrightarrow{\text{Transpeptidase}} \text{Peptide}-\overset{O}{\overset{\|}{C}}-\text{NHR}' + H_2NR$$

• **Figure 8.12** (a) Nucleophilic addition–elimination leading to peptide hydrolysis and mediated by a peptidase. (b) The same process causing transfer of an acyl group to a second amine by a transpeptidase.

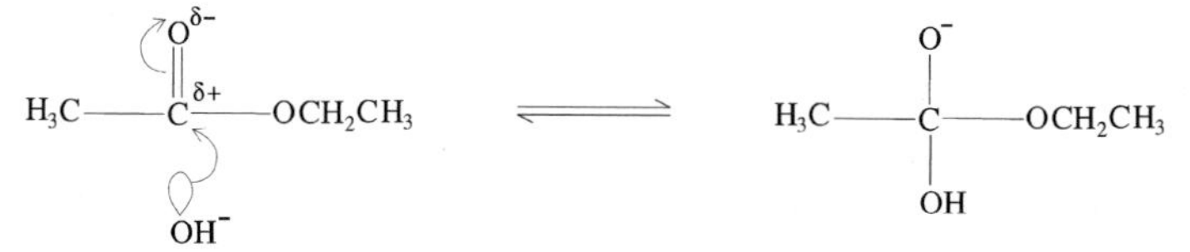

• **Figure 8.13** Nucleophilic addition of hydroxide ion to a polar double bond in an ester, ethyl ethanoate, followed by elimination to give ethanol and the ethanoate ion.

Lone pair electrons on the hydroxide ion move to the slightly positive carbonyl carbon atom, while electrons in the double bond shift to the slightly negative oxygen atom

A new carbon–oxygen bond is formed, while the carbonyl oxygen atom has gained a negative charge

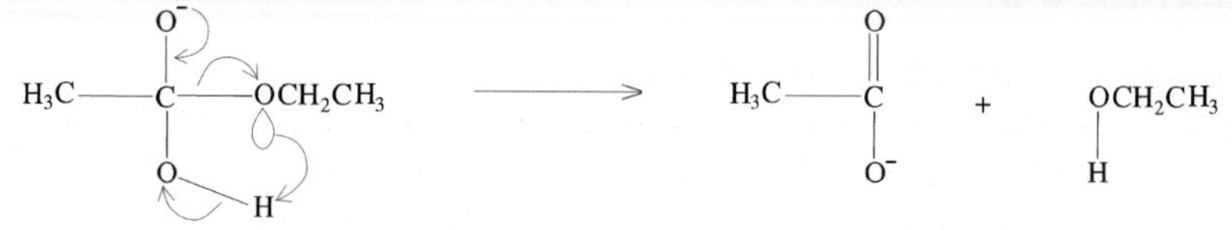

The anion intermediate rearranges, electrons shift from the negative oxygen to reform the carbonyl double bond. A proton moves from one oxygen atom to another

Ethanoate ion is formed Ethanol is released

• **Figure 8.14** A key stage in the hydrolysis of a peptide adjacent to an aromatic amino acid catalysed by the serine protease, chymotrypsin.

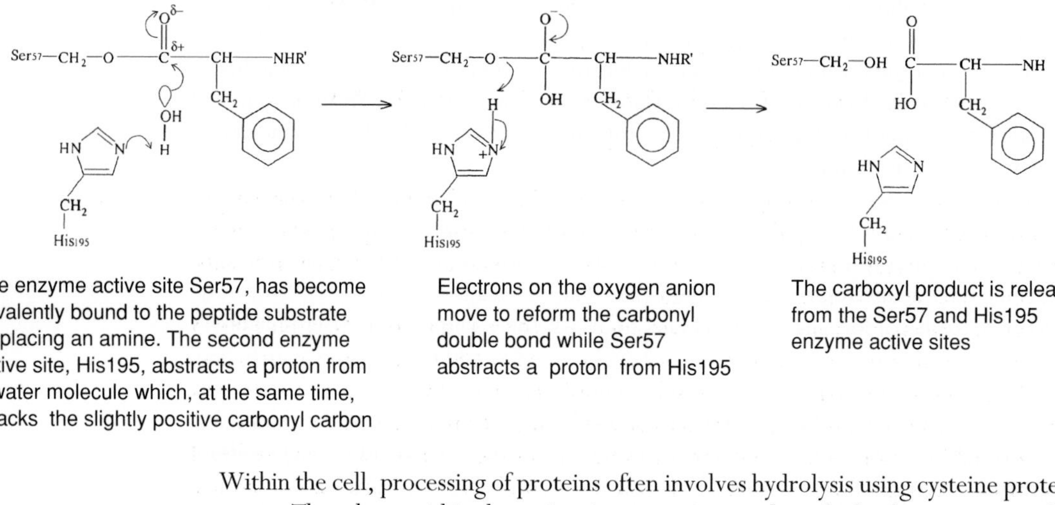

The enzyme active site Ser57, has become covalently bound to the peptide substrate displacing an amine. The second enzyme active site, His195, abstracts a proton from a water molecule which, at the same time, attacks the slightly positive carbonyl carbon

Electrons on the oxygen anion move to reform the carbonyl double bond while Ser57 abstracts a proton from His195

The carboxyl product is released from the Ser57 and His195 enzyme active sites

Within the cell, processing of proteins often involves hydrolysis using cysteine protease enzymes. These have within the active site a cysteine residue which takes part in covalent catalysis. The mechanism requires the cysteine residue to act as a nucleophile and attack the electropositive carbon atom in the carbonyl group of a peptide. An intermediate thioester is formed which is in turn hydrolysed with water acting as a nucleophile to release the amine and carboxylic acid and reform the original cysteine residue. An example of the mechanism is provided by papain, an endoprotease enzyme, which cleaves peptide linkages preferentially at basic amino acid sites, such as arginine and lysine. The reaction is summarised in Figure 8.15.

QUESTION 8.5

The reaction between a nucleophile and a polar double bond often involves addition followed by elimination. Use a suitable example to illustrate the mechanism of the reaction.

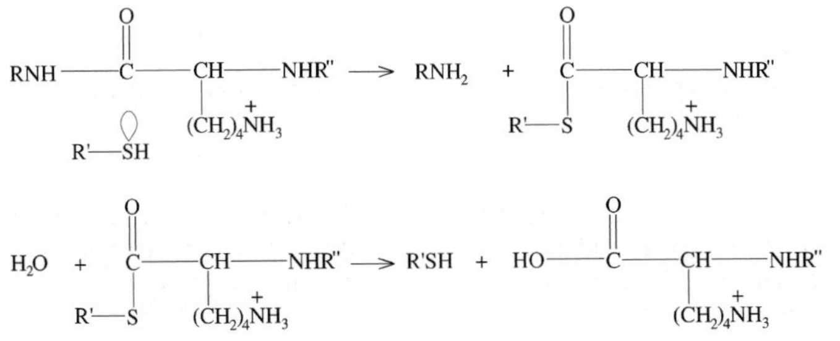

$$RNH-C(=O)(\cdots SH-R')-CH(CH_2)_4\overset{+}{N}H_3-NHR'' \longrightarrow RNH_2 + R'-S-C(=O)-CH(CH_2)_4\overset{+}{N}H_3-NHR''$$

$$H_2O + R'-S-C(=O)-CH(CH_2)_4\overset{+}{N}H_3-NHR'' \longrightarrow R'SH + HO-C(=O)-CH(CH_2)_4\overset{+}{N}H_3-NHR''$$

• **Figure 8.15** Two successive reactions by nucleophilic groups (R'SH and H_2O) linked to enzymes in the hydrolysis of a peptide adjacent to a basic amino acid residue. The enzyme involved is papain.

■ 8.8 FREE RADICAL REACTIONS

Oxidation by oxygen from air is usually a free radical process

Free radicals are often formed by the energy in sunlight or ultraviolet light interacting with simple organic molecules. The atmosphere contains small traces of freons, which are compounds containing chlorine and fluorine atoms linked to carbon. They are present through their use as propellant gases in aerosol cans or as refrigerant fluids. These compounds are decomposed by sunlight to give chlorine radicals. The radicals formed react with ozone, O_3, in the upper atmosphere to form oxygen molecules, O_2, and oxygen atoms, O. The mechanism of the reaction can be represented in the following way:

chloropentafluoroethane → pentafluoroethyl radical + chlorine radical

$$CF_3CF_2Cl \rightarrow CF_3CF_2^{\cdot} + Cl^{\cdot}$$

ozone + chlorine radical → oxygen molecule + chloroxy radical

$$O_3 + Cl^{\cdot} \rightarrow O_2 + ClO^{\cdot}$$

This process depletes the all-important ozone layer, allowing harmful short-wavelength ultraviolet radiation to reach the surface of the earth. The reaction is particularly significant as the mechanism allows the chloroxy radicals formed to react further in a chain reaction and destroy fresh ozone molecules.

When the electron structure of the oxygen molecule was discussed (section 2.8), it was concluded that two unpaired electrons are present in π antibonding orbitals. Oxygen, O_2, is a powerful oxidising agent in biological systems. It may react by receiving a pair of electrons from a substrate that is to be oxidised or it can receive a single electron from a suitable donor. A substance that can behave in the second way is the reduced form of the cofactor flavin, $FADH_2$. When it loses an electron, it forms a free radical intermediate, flavin semiquinone, $FADH^{\cdot}$, which is converted to the oxidised form of flavin, FAD, by losing a second unpaired electron (Figure 8.16). The oxygen molecule is reduced to hydrogen peroxide, H_2O_2. A **free radical** is a species which contains an unpaired electron, usually on a carbon atom. Free radicals are highly reactive and combine with a suitable donor of an unpaired electron to form a conventional electron pair.

FAD acts as a cofactor with the enzyme monoamine oxidase in the free radical oxidation of amines to aldehydes. The mechanism involves two, one-electron transfer steps from FAD to form an imine which subsequently undergoes hydrolysis (Figure 8.17). Alternatively, the mechanism may involve carbanion and carbon free radical intermediates. A similar radical mechanism can be invoked for the oxidation of succinate to fumarate

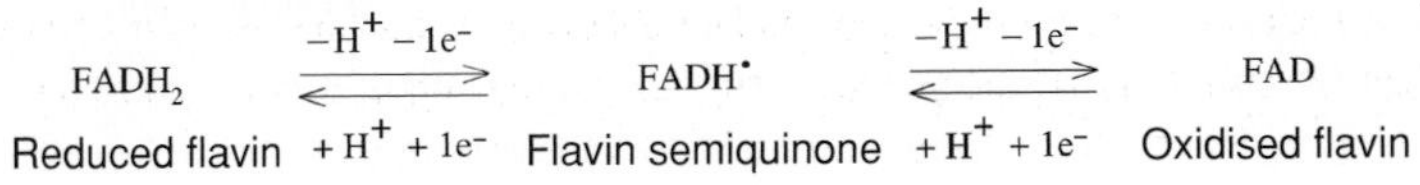

$$\underset{\text{Reduced flavin}}{FADH_2} \underset{+H^+ + 1e^-}{\overset{-H^+ - 1e^-}{\rightleftharpoons}} \underset{\text{Flavin semiquinone}}{FADH^{\cdot}} \underset{+H^+ + 1e^-}{\overset{-H^+ - 1e^-}{\rightleftharpoons}} \underset{\text{Oxidised flavin}}{FAD}$$

• **Figure 8.16** The three redox states of riboflavin.

• **Figure 8.17** Free radical oxidation of an amine to an aldehyde mediated by FAD and monoamine oxidase.

$$RCH_2CH_2NH_2 \xrightarrow[FAD \quad FADH^{\bullet}]{} RCH_2CH_2NH_2^{+\bullet} \xrightarrow[FADH^{\bullet} \quad FADH_2]{} RCH_2CH{=}NH_2^+ \xrightarrow{H_2O} RCH_2CHO + NH_4^+$$

• **Figure 8.18** Oxidation of succinate to fumarate using succinate dehydrogenase in the presence of FAD. A free radical mechanism is involved.

$$\underset{\text{Succinate}}{^-O_2CCH_2CH_2CO_2^-} \xrightarrow[FAD \quad FADH_2]{} \underset{\text{Fumarate}}{^-O_2CCH{=}CHCO_2^-}$$

in the presence of succinate dehydrogenase (Figure 8.18). Amino acids are subject to oxidation forming 2-ketocarboxylic acids using the enzyme D-amino acid oxidase in the presence of FAD.

The woody tissues in plants contain significant proportions of lignin. Around one-third of the dry weight of wood is lignin. This substance is a complex aromatic material containing phenylpropanoid residues linked together in a heterogeneous manner. A phenylpropanoid group contains three carbon atoms joined to a benzene ring. One of the biosynthetic precursors of lignin is coniferyl alcohol (Figure 8.19). Free radical coupling of coniferyl alcohol, or the related precursors, takes place to give any one of several possible single or multiple links between the molecules to form the giant, crosslinked matrix that is lignin. The radical coupling process is initiated by formation of a phenoxy radical (Figure 8.19). This occurs under the influence of one of the peroxidase, phenolase or tyrosinase enzymes that occur widely in plants. Once a radical is formed by enzyme mediation, the subsequent radical polymerisation occurs by a simple chemical reaction (Figure 8.19).

QUESTION 8.6

Flavin can exist in three redox states.

(a) Which of the three states contains a free radical?

(b) Explain how this observation is significant in biological oxidation processes involving molecular oxygen.

8.9 CARBON–CARBON BOND FORMATION IN BIOSYNTHESIS

Carbon–carbon bond formation is a key step in biosynthesis

An essential step in intracellular processes is the formation of a carbon–carbon bond. Biosynthesis is based on the formation of carbon compounds in the cell by such steps.

• **Figure 8.19** Simplified outline of radical formation from coniferyl alcohol leading to lignin biosynthesis.

$$\underset{\text{Coniferyl alcohol}}{HO{-}C_6H_3(OCH_3){-}CH{=}CHCH_2OH} \xrightarrow{\text{Peroxidase}} \underset{\text{Intermediate phenoxy free radical}}{^{\bullet}O{-}C_6H_3(OCH_3){-}CH{=}CHCH_2OH} \xrightarrow[\text{polymerisation}]{\text{Radical}} \text{Lignin}$$

• **Figure 8.20** Mechanism of carbon–carbon bond formation in the aldol reaction with propanone.

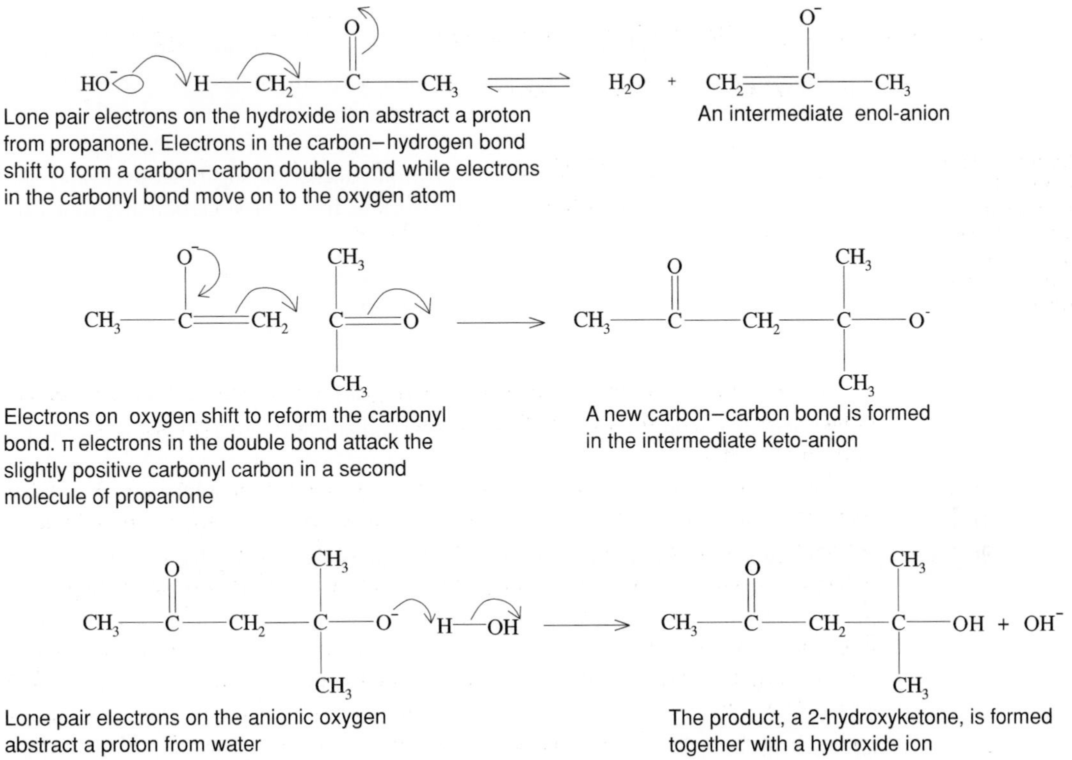

HO^{-} H—CH_2—C(=O)—CH_3 ⇌ H_2O + CH_2=C(O^{-})—CH_3

Lone pair electrons on the hydroxide ion abstract a proton from propanone. Electrons in the carbon–hydrogen bond shift to form a carbon–carbon double bond while electrons in the carbonyl bond move on to the oxygen atom

An intermediate enol-anion

CH_3—C(O^{-})=CH_2 + (CH_3)$_2$C=O ⟶ CH_3—C(=O)—CH_2—C(CH_3)$_2$—O^{-}

Electrons on oxygen shift to reform the carbonyl bond. π electrons in the double bond attack the slightly positive carbonyl carbon in a second molecule of propanone

A new carbon–carbon bond is formed in the intermediate keto-anion

CH_3—C(=O)—CH_2—C(CH_3)$_2$—O^{-} H—OH ⟶ CH_3—C(=O)—CH_2—C(CH_3)$_2$—OH + OH^{-}

Lone pair electrons on the anionic oxygen abstract a proton from water

The product, a 2-hydroxyketone, is formed together with a hydroxide ion

The products formed are usually primary metabolites such as carbohydrates and amino acids which are required to support the function of the cell and the life of the organism. In turn, secondary metabolites may be synthesised. These are valuable to the cell but not essential to its survival.

Carbon–carbon bond formation can occur in one of three ways:

- nucleophilic attack by a carbanion at the positive, carbon end of a polar carbonyl bond,
- electrophilic attack by a carbonium ion (carbocation) at the negative charge cloud of a carbon–carbon double bond (alkene),
- attack by a carbon-centred free radical on a carbon–carbon double bond (alkene) or by combination of two phenoxy free radicals.

The carbanion route appears to dominate biosynthetic mechanisms. It requires a pair of electrons to be donated by the carbanion (a Lewis base) to a suitable carbon acceptor (a Lewis acid). An example is provided in organic chemistry by the aldol condensation of propanone (acetone) with a second propanone molecule which occurs in aqueous alkali. The reaction mechanism is shown in Figure 8.20. In the first stage, a hydroxide ion abstracts a proton from propanone to form a carbanion. This uses π electrons in the double bond to attack a slightly positive carbonyl carbon in the second molecule to form a new carbon–carbon bond. The anion generated in this way is protonated by water to give the hydroxy-ketone product and reform the hydroxide ion. Within the cell, the harsh conditions necessary for the formation of simple carbanions do not occur. Thus enzymes are required to intervene and stabilise the carbanion intermediates by dispersing the negative charge over several atoms. Some examples of biosynthetic reactions involving aldol-type processes or related carbanion reactions will be outlined.

Dihydroxyacetone phosphate undergoes a reversible reaction with glyceraldehyde-3-phosphate to form fructose-1,6-biphosphate. The enzyme involved is fructose-1,6-biphosphate aldolase. The reaction is a major step in the Calvin cycle which is itself part of the processes making up photosynthesis. In animals, fructose-1,6-biphosphate is implicated in glycolysis, the catabolism of monosaccharide sugars to pyruvate, and in gluconeogenesis, the biosynthesis of carbohydrates from three-carbon and four-carbon precursors.

The dihydroxycarboxylic acid, mevalonic acid, with six carbon atoms, is formed by an initial condensation of two acetyl coenzyme A (acetyl CoA) units to give the four-carbon intermediate, 3-ketobutyryl CoA. This is followed by reaction with a third two-carbon unit in the form of acetyl CoA and subsequent reduction to give mevalonic acid. Each of the three steps is catalysed by an enzyme. Mevalonic acid is significant in acting as a precursor for the biosynthesis of terpenes and steroids. Acetyl CoA is also involved in the synthesis of fatty acids. These are long-chain carboxylic acids, usually with 12 to 20 carbon atoms (section 7.2). They are useful as a store of metabolic energy when they are converted to the corresponding triesters with glycerol (fats). The biosynthetic sequence involves the addition of two-carbon units to an acyl chain bound to an acyl carrier protein, ACP, by a thioester link (section 9.4). Initially, acetyl CoA transfers a two-carbon fragment to ACP. In subsequent stages malonyl CoA transfers a malonyl, three-carbon unit to ACP which undergoes decarboxylation – loss of carbon dioxide – in a concerted mechanism. Each stage is thus able to add a further two carbon atoms.

A major role of carbonium ion intermediates in biosynthesis is in the conversion of allyl pyrophosphates into terpenes. A wide range of terpenes are important in plants, while in animals hormones and steroid lipids are major compounds within this group. Since five carbon atoms are present in the initial dimethylallylpyrophosphate, DMAPP, then the terpenes formed contain multiples of five carbon atoms. The terpenes camphor, menthol, geraniol and pinene each contain 10 carbon atoms while pentalenene, a precursor of pentalenolactone antibiotics, has 15 carbon atoms. The first stage in the synthesis of a terpene from DMAPP is the loss of a pyrophosphate group to leave a carbonium ion. This is then attacked by the π electrons in the double bond of a second molecule of DMAPP to form a new carbon–carbon bond.

Radical intermediates in biosynthesis are involved principally in the formation of the woody plant material lignin (section 8.7). The combination of two radicals each with one unpaired electron results in the formation of a new, two-electron, carbon–carbon bond. Several different types of carbon–carbon bond are formed in the biosynthesis of lignin. These bonds are characterised by the common feature of having their origin in a radical coupling process.

■ SUMMARY

The course followed by a reaction in organic chemistry can be understood by a consideration of the reaction mechanism. Within an organic molecule, reactive sites can be identified and recognised in terms of functional groups and bond characteristics. A small number of mechanistic processes account for most of the reactions that occur. One functional group may replace another when a nucleophile is attracted to a positive centre in a substrate; this is nucleophilic substitution. The π electrons in the double bond of an alkene may attack an electrophile allowing addition to the double bond to occur. The mechanism is called electrophilic addition to a non-polar double bond. Under the appropriate conditions the reaction can be reversed. The loss of two groups from a reactant leads to the formation of a new alkene; this is elimination. A carbonyl compound contains a polar double bond, the positive end of which can attract a nucleophile, allowing an addition to take place. This may be followed by elimination to reform a double bond and give a product. The overall process is called addition–elimination.

In all of the mechanisms mentioned, the electrons involved in forming or breaking chemical bonds move in pairs. In some alternative reaction mechanisms the electrons move singly. These reactions often involve oxygen and are called free radical reactions. Each of these mechanistic processes has a parallel in the living world. Reactions within the cell can be interpreted in terms of these organic schemes. This leads to a better understanding of the pathways involved in anabolic and catabolic processes. However, there are important differences. Organic chemistry is concerned with powerful reagents and harsh conditions applied to the reactions of relatively simple reactants. In metabolism, reactions occur under mild conditions at physiological pH and temperature. The reactants are diverse, they may be small or very large molecules and may not be amenable to facile chemical change. These difficulties are overcome by the intercession of specific, powerful enzyme catalysts which mediate and control the reaction mechanisms.

FURTHER READING

Bugg, T. (1997) *An Introduction to Enzyme Chemistry*. Blackwell Science, London. (A clear, readable sourcebook for the mechanisms of enzyme biochemistry.)

Fersht, A. (1985) *Enzyme Structure and Mechanism*, 2nd edn. Freeman, Oxford. (A useful and detailed account of mechanism with the emphasis on the practical approach.)

END OF CHAPTER QUESTIONS

Question 8.7

(a) What is meant by the term reactive site?

(b) Use the examples (i) to (iii) given below to explain how Lewis acid and Lewis base reactive sites may occur within molecules.

(i) Butanone, $CH_3CH_2COCH_3$

(ii) 2-Aminopropane, $CH_3CH(NH_2)CH_3$

(iii) Methyl ethanoate, CH_3COOCH_3

Question 8.8

The two mechanistic processes of nucleophilic substitution and elimination may each occur by a concerted mechanism. Select from this chapter an appropriate example to illustrate this mechanistic concept.

Question 8.9

'Elimination to form an alkene double bond is often linked with addition to a double bond in reactions within the cell.' Use an example from the citric acid cycle to justify this statement.

Question 8.10

(a) Identify two important mechanistic routes that are implicated in the formation of carbon–carbon bonds in biosynthesis.

(b) Describe an example of one of these routes and explain its significance.

9 SULPHUR AND PHOSPHORUS

9.1 INTRODUCTION

The mineral elements sulphur and phosphorus are essential for all living organisms. Sulphur can be used as an alternative to oxygen by some bacteria. Sulphur is normally taken in by bacteria and plants as sulphate where it is reduced before being incorporated into proteins and coenzymes. Sulphur, in the form of the thiol group, is used as a redox mediator and facilitator of chemical reactions as well as a means of providing covalent bonds between polypeptides in protein conformation.

Phosphorus is mainly found in living organisms as phosphate salts and the related compounds phosphate and polyphosphate esters. The role of 'phosphates' in living cells is diverse. Protein phosphate esters can help to regulate metabolic activity and mediate hormonal events in cells. Sugar phosphate esters are used in activating sugars in metabolic breakdown and are components of nucleotides. Nucleotides, such as adenosine-5′-triphosphate (ATP), are involved in energy storage and transfer. Phosphodiesters are used to link nucleotides together when DNA or RNA is polymerised and form a significant structural consideration in DNA.

9.2 THE ELECTRON SHELL STRUCTURE AND VALENCY OF PHOSPHORUS AND SULPHUR

$^{14}_{7}$N 14.0	$^{16}_{8}$O 16.0
$^{31}_{15}$P 31.0	$^{32}_{16}$S 32.1

A section of the periodic table (see Table 1.4 and Figure 1.1) for phosphorus and sulphur is shown in the margin. The elements phosphorus and sulphur have atomic numbers 15 and 16, respectively, and thus have 15 or 16 electrons in their electron shells. The relative position in the periodic table of both elements is in the third row directly below nitrogen and oxygen, respectively. Thus it is expected that phosphorus and sulphur will share properties in common with nitrogen and oxygen, respectively. Examination of Table 2.1 illustrates that there are similarities and differences in the valency numbers of elements in the same column of the periodic table. Thus nitrogen and phosphorus have a valency of three whilst phosphorus has an additional valency of five. Oxygen and sulphur have valencies of two whilst sulphur has additional valencies of four and six. It will be helpful to examine the orbital arrangement of these atoms if we are to understand the similarities and differences between the atoms in the second and third rows of the periodic table. Table 1.5 suggests that the differences between second- and third-row elements is due to the presence of a d set of orbitals in the third valence shell. We can write the electron structure of the valence shell of phosphorus and sulphur as (see section 1.7):

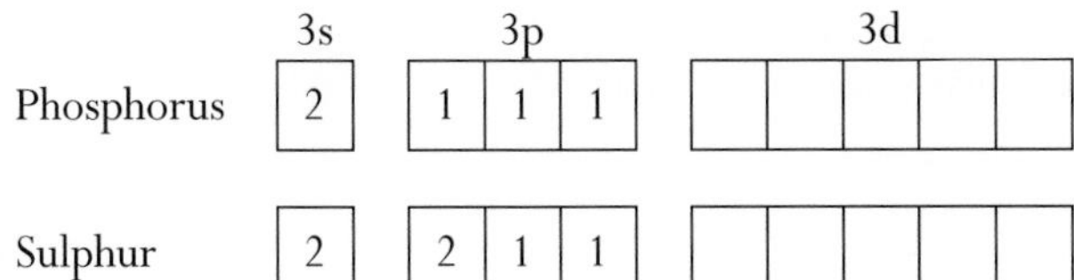

The valencies of three for phosphorus and two for sulphur derive from the number of electrons required to complete the 3p orbital shells of each element. The valency of five for phosphorus is a result of unpairing of the electrons in the 3s orbital to give the arrangement:

	3s	3p			3d				
Phosphorus	1	1	1	1	1				

There are now five orbitals in the third shell of electrons containing one electron. Filling of these orbitals by covalent bond formation produces a stable structure for the phosphorus atom. The valency of four for sulphur is derived from unpairing of the filled 3p orbital to give:

The variable valency of sulphur and phosphorus is due to unpairing of electrons in 3s and/or 3p filled orbitals

	3s	3p			3d				
Sulphur	2	1	1	1	1				

with four singly filled orbitals. The valency of six comes from unpairing of both the 3s and 3p orbitals to yield:

	3s	3p			3d				
Sulphur	1	1	1	1	1	1			

with six singly filled orbitals.

The valencies most commonly found in biological systems for phosphorus are five whilst sulphur has two, four and six.

WORKED EXAMPLE 9.1

(a) Show the structure of sulphur dioxide.

(b) What is the valency of sulphur in sulphur dioxide?

(c) Which, if any, of the sulphur orbitals have been unpaired?

ANSWER

(a) The structure of sulphur dioxide is O=S=O.

(b) The valency of sulphur in sulphur dioxide is four. This can be worked out by counting the bonds attached to the sulphur atom.

(c) The valency of four for sulphur is derived from unpairing of the filled 3p orbital.

QUESTION 9.1

(a) Show the structure of sulphur trioxide.

(b) What is the valency of sulphur in sulphur trioxide?

(c) Which, if any, of the sulphur orbitals have been unpaired?

9.3 SULPHUR

Atmospheric sulphur contributes to acid rain

The source of sulphur for most plants and microbes is as **oxy** anions of sulphur. The two main oxy anions are sulphite and sulphate (often added to crops as the fertiliser ammonium sulphate). The sulphur in sulphite, valency four, is formed from the solubilisation of atmospheric sulphur dioxide in water to form sulphurous acid:

$$SO_2 + H_2O \rightleftharpoons H_2SO_3$$

Sulphur dioxide is a by-product of many chemical processes and sulphurous acid is one of the components of acid rain. In solution the acid readily dissociates:

$$H_2SO_3 \rightleftharpoons H^+ + HSO_3^- \rightleftharpoons 2H^+ + SO_3^{2-}$$

with a pK_1 of 1.8 and a pK_2 of 6.92 at 25°C.
Sulphur dioxide in the atmosphere is capable of being oxidised by oxygen to form sulphur trioxide:

$$2SO_2 + O_2 \rightleftharpoons 2SO_3 \textit{ or } SO_2 + O_3 \rightleftharpoons SO_3 + O_2$$

Sulphur trioxide, in which sulphur has a valency of six, is very soluble in water, producing sulphuric acid:

$$SO_3 + H_2O \rightleftharpoons H_2SO_4$$

Sulphuric acid is widely used in chemical manufacture although its anion in solution, sulphate, is of more biological importance. In solution the acid readily dissociates:

$$H_2SO_4 \rightleftharpoons H^+ + HSO_4^- \rightleftharpoons 2H^+ + SO_4^{2-}$$

Sulphate esters help to form protein sugar interactions in glycosaminoglycans

H_2SO_4 is a strong acid, whilst HSO_4^- dissociation has a pK_2 of 1.92 and is a weak acid. The sulphate ion can form sulphate esters in the following reaction:

$$ROH + SO_4^{2-} \rightleftharpoons ROSO_3^- + OH^-$$

Sulphate esters are components in many body glycosaminoglycans where the negative charge on the sulphate can bind to appropriate groups in proteins to form strong ionic interactions. Some of the sugar sulphate components of glycosaminoglycans such as chondroitin sulphate, keratan sulphate and heparin are shown in Table 9.1.

Many other compounds of biological interest are shown in Table 9.1. The sulphur in the compounds shown, apart from the oxy anions, always has a valency of two. The sulphur absorbed by plants and microbes is normally as sulphite, valency four, or sulphate, valency six. Sulphur must be reduced by the organisms to give it a valency of two so that

Table 9.1 Some biological molecules containing sulphur

Name of sulphur-containing compound	*Example structure (some hydrogens omitted)*	*Biological function*
Biotin	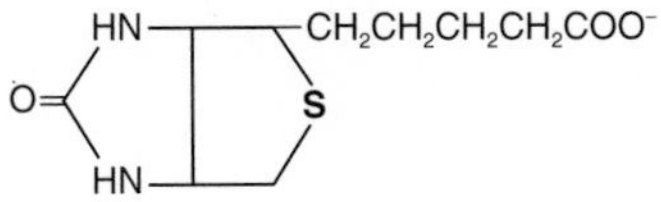 HN, O, HN, **S**, $CH_2CH_2CH_2CH_2COO^-$	Involved in carboxylation reactions
Chondroitin sulphate	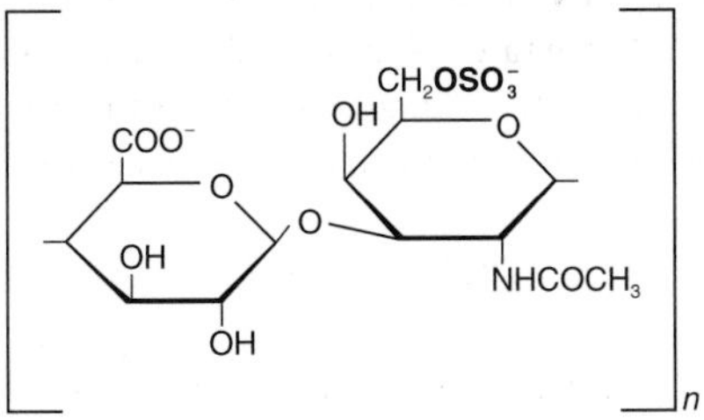 [COO^-, O, OH, OH, O, CH_2**OSO_3^-**, OH, O, $NHCOCH_3$]$_n$	Cartilage component
Coenzyme A	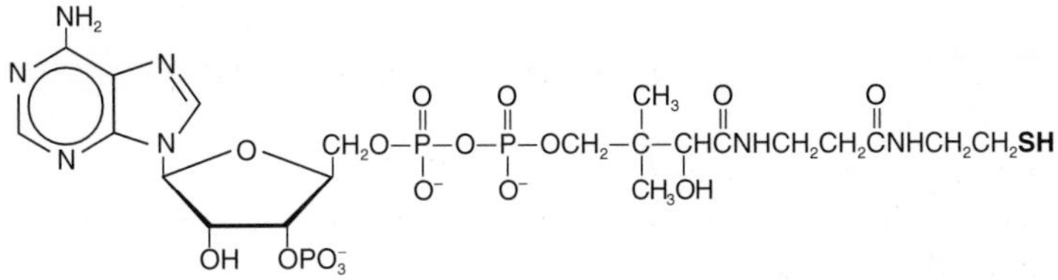 NH_2, N, N, N, N, O, OH, OPO_3^-, $CH_2O-P(=O)(O^-)-O-P(=O)(O^-)-OCH_2-C(CH_3)(CH_3)-CH(OH)CNHCH_2CH_2CNHCH_2CH_2$**SH**	Wide use in biological catalysis
Cysteine	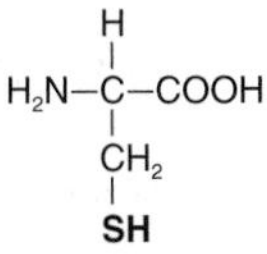 $H_2N-C(H)(CH_2$**SH**$)-COOH$	Amino acid
Glutathione	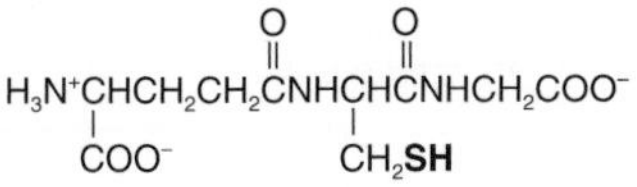 $H_3N^+CH(COO^-)CH_2CH_2C(=O)NHCH(CH_2$**SH**$)C(=O)NHCH_2COO^-$	Biological redox mediator
Heparin sulphate	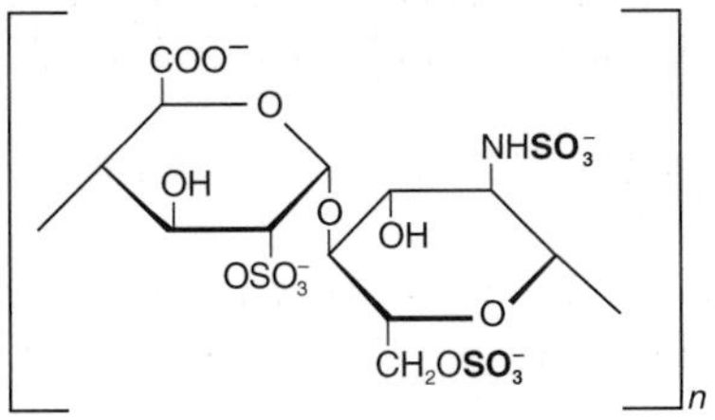 [COO^-, O, OH, OSO_3^-, O, NH**SO_3^-**, OH, O, CH_2O**SO_3^-**]$_n$	Anticoagulant
Keratan sulphate	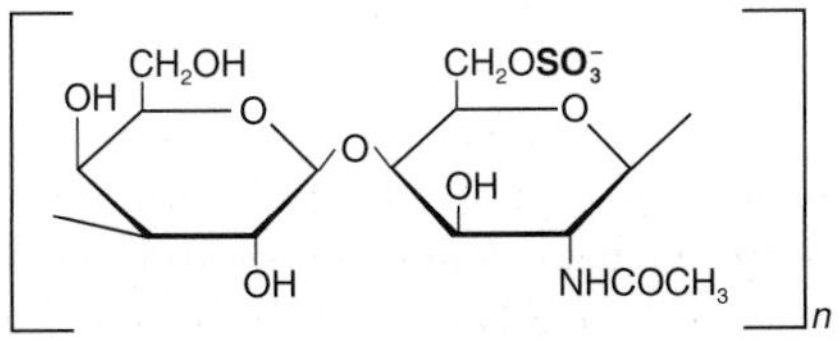 [CH_2OH, OH, O, OH, O, CH_2O**SO_3^-**, OH, O, $NHCOCH_3$]$_n$	Cartilage component
Lipoic acid	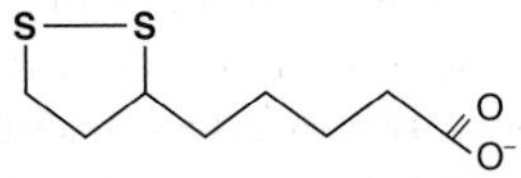 **S**—**S**, O, O^-	Fatty acid metabolism

Table 9.1 (cont'd)

Name of sulphur-containing compound	*Example structure (some hydrogens omitted)*	*Biological function*
Methionine	$CH_3\mathbf{S}CH_2CH_2CH(NH_3^+)COO^-$	Amino acid
S-adenosylmethionine	adenine–(ribose ring: O; OH, OH)–$CH_3\mathbf{S}CH_2CH_2CH(NH_3^+)COO^-$	Methyl group transfer
Sulphanilamide	H_2N–C_6H_4–$\mathbf{SO_2}NH_2$	Antibiotic
Sulphate	SO_4^{2-}	Fertiliser
Sulphite	SO_3^{2-}	Pollutant, bacteriocide
Thiamine	pyrimidine ring (NH_2, N, N, H_3C)–CH_2–N^+ thiazole ring (**S**, CH_3, CH_2CH_2OH)	Vitamin used in biological catalysis
Thioguanine	purine ring (**SH**, N, N, N, NH, H_2N)	Artificial growth regulator

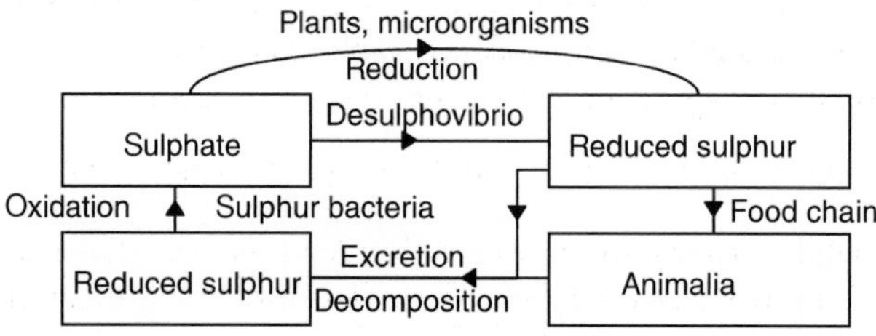

• **Figure 9.1**
The sulphur cycle.

it can be incorporated into the biomolecules shown in Table 9.1. In sulphate reduction the sulphate is attached to a biological carrier and reduced first to sulphite and then to hydrogen sulphide:

$$\text{carrier} + SO_4^{2-} \rightarrow \text{carrier-}SO_4^- + \text{electrons} + \text{protons} \rightarrow SO_3^{2-} + H_2O + \text{carrier}$$

$$SO_3^{2-} + \text{electrons} + \text{protons} \rightarrow H_2S$$

The sulphate reduction involves electrons donated by thioredoxin, whilst sulphite is reduced by NADPH to form water and sulphide.
The overall assimilation of sulphur is part of a sulphur cycle (see Figure 9.1) which has similarities to the carbon cycle shown in Figure 6.1.

WORKED EXAMPLE 9.2

Explain why sulphur needs to be reduced in its assimilation into living organisms.

ANSWER

Sulphur is normally absorbed as sulphate, a highly oxidised form of sulphur. Most biological compounds use sulphur in a reduced form, such as hydrogen sulphide, for synthesis of biomolecules. Thus reduction of sulphur must precede its assimilation.

QUESTION 9.2

(a) What is the valency of sulphur in coenzyme A?

(b) How does this differ from the valency of sulphur in the sulphur source, sulphate?

9.4 THE THIOL GROUP AND THIOL ESTERS

Thiols have the general formula RSH where —SH is called the thiol group. The structures of methanol and methanethiol are shown in the margin. The hydroxyl group and thiol group have similarities and differences as might be expected from atoms in the same column of the periodic table (see section 9.1). The hydroxyl oxygen and thiol sulphur have a valency of two and adopt a similar shape in solution. Table 3.2 shows that oxygen (3.5) is electronegative with respect to carbon (2.5) whilst sulphur (2.5) and carbon have the same electronegativity. Thus alcohols but not the thiols are capable of forming intermolecular hydrogen bonds in solution. Consequently, thiols are more volatile in solution than the corresponding alcohol, despite the increased mass of the sulphur atom. The unpleasant smell of many sulphur-containing compounds – hydrogen sulphide, for example, has the well-known smell of rotten eggs – is partly a result of their volatility. The sulphur atom is much larger than the oxygen atom as the atomic radius of the third-level valence shell is much greater than the second-level shell. The sulphur–carbon bond is weaker (259 kJ mol^{-1}) than the carbon–carbon bond (348 kJ mol^{-1}) because of the mismatch in atomic size between second and third valence shells of electrons.

C O

C S

Comparison of methanol (above) and methanethiol (below)

Two thiol groups are capable of oxidation to form a covalent bond:

$$\mathrm{R{-}S{-}H + R'{-}S{-}H \rightleftharpoons R{-}S{-}S{-}R' + 2H^+ + 2e^-}$$

where the two sulphur atoms are covalently bonded together. The resulting group is called a disulphide bridge. The disulphide bridge structure is important in the formation of cystine covalent bonds between and within polypeptide chains to hold proteins in their three-dimensional conformation. The biological activity of some enzymes, such as the ATP-generating enzyme found in chloroplasts, can be switched on or off by oxidation or reduction of the disulphide bridge. The ability of thiol groups to be reversibly oxidised is measured by the standard reduction potential (see Table 10.1). Typical values range from −0.23 V to −0.29 V, meaning that such compounds fall into the middle of the range of biological oxidation–reduction reactions and so can act as oxidant or reductant. The molecule glutathione (see Table 9.1) is involved in keeping the contents of biological cells in the correct oxidation state. It is involved in the repair of oxidised disulphide bridges in proteins, as shown in Figure 9.2. Glutathione is also one of the molecules responsible for removing dangerous and reactive peroxides (R—O—O—H) that are formed as the result of cellular oxidation reactions:

$$\mathrm{2\ glutathione{-}SH + ROOH \rightleftharpoons glutathione{-}S{-}S{-}glutathione + ROH + H_2O}$$

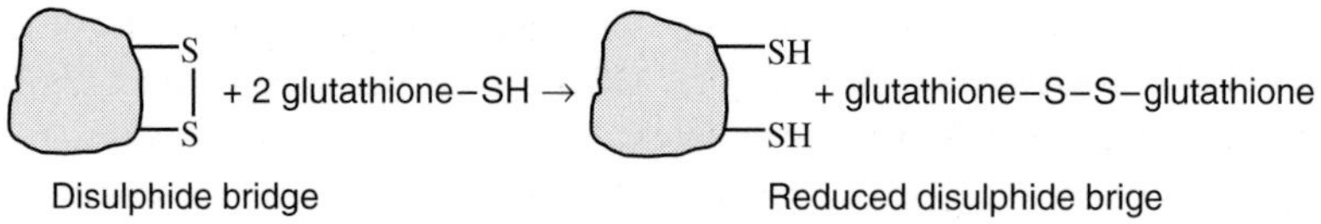

• **Figure 9.2** Reduction of unwanted disulphide bridges in proteins by glutathione.

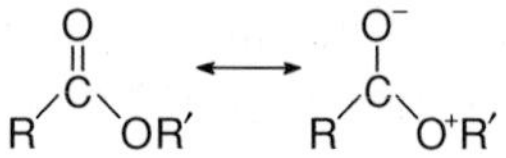

+ 2 glutathione–SH → + glutathione–S–S–glutathione

Disulphide bridge — Reduced disulphide brige

• **Figure 9.3** Resonance stabilisation of organic esters.

This produces alcohols, which are less reactive and so less dangerous. Glutathione is reduced enzymatically to keep the cellular ratio of reduced to oxidised at 500:1.

Proximal thiol groups are also found in lipoic acid (see Table 9.1). The two thiol groups help carry out some of the reactions that are involved in the oxidative decarboxylation of some β-keto acids of fatty acids to release cellular energy.

Thiols are capable of reacting with organic acids to form thioesters:

$$\underset{\text{thiol}}{RSH} + \underset{\text{organic acid}}{HOOCR'} \rightleftharpoons \underset{\text{thioester}}{RSCOR'} + \underset{\text{water}}{H_2O}$$

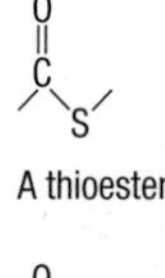

A thioester

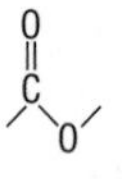

An ester

The structure of a thioester is compared with an organic ester in the margin. Thioesters are more reactive than carboxylic acid esters because organic esters are stabilised by resonance hybridisation (Figure 9.3) which does not happen for thioesters. This, coupled with the fact that the carbon–sulphur bond in thioesters is weaker than the carbon–oxygen bond in carboxylic acid esters, means that living cells have evolved the use of thioesters as intermediates in chemical reactions.

This means that thiols such as coenzyme A (see Table 9.1) are used to form thioesters as intermediates in many biochemical reactions.

QUESTION 9.3

(a) Explain two biological roles of the thiol group in reduction–oxidation reactions.

(b) Explain why thioesters are used as intermediates in biochemical reactions rather than organic esters.

■ 9.5 PHOSPHATE, PYROPHOSPHATE AND POLYPHOSPHATES

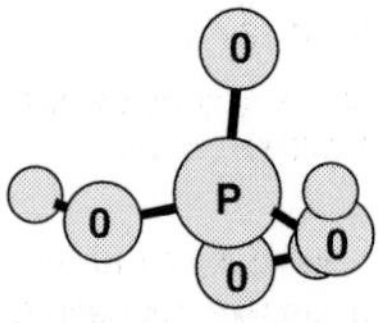

Most of the phosphorus in the biosphere exists as the phosphate anion. The normal form of the phosphate anion is based on a structure of PO_4^{3-} where the phosphorus atom has a valency of five. The four oxygen atoms are arranged tetrahedrally around the phosphorus with only one of the oxygens double bonded to the phosphorus atom. The electrons in the double bond are delocalised equally between all four oxygen atoms to give this symmetrical arrangement for the four oxygen atoms.

Tetrahedral arrangement of oxygens in phosphoric acid

The three charges on the phosphate are capable of protonation to form phosphoric acid, as shown in the diagram in the margin. The important properties of phosphate relate to its acidic behaviour. Phosphoric acid is capable of dissociating the hydrogens in three successive steps:

with a pK_1 of 2.2, pK_2 of 7.2 and pK_3 of 12.4. Thus at physiological pHs most of the phosphate exists as a mixture of $H_2PO_4^-$ and HPO_4^{2-}. The neutral pK_2 makes phosphate a useful buffer for many biological applications.

The phosphate ion is highly polar and readily soluble in water. The negative charges are capable of forming salt links when incorporated into proteins. Calcium phosphate salts are extremely important in biological systems. One of the features of many calcium phosphate salts is their insolubility whilst another is the ability to form crystals of great strength. The salt hydroxyapatite, $Ca_{10}(PO_4)_6(OH)_2$, is one of the major components of bones, teeth and enamel, where it forms rod-shaped crystals and is associated with the protein collagen. Crystals of calcium and magnesium phosphate comprise one of the types of renal stones.

Two phosphates can be reacted together to form pyrophosphate where there is an oxygen bridging the two phosphorus atoms, as shown in Figure 9.4 for the acid forms. Bridging the two phosphates results in loss of two ionisable groups; the pyrophosphate valency is four compared with six for the two phosphates. The oxygen linking the two phosphorus atoms is called a **phosphoanhydride** bond. A further phosphate can be similarly added to pyrophosphate to form the triphosphate ion. Such 'polyphosphates' and their esters are important reactive groups within biological compounds. They are used in phosphate-rich fertilisers such as superphosphate. Pyrophosphate is also used as a buffering agent in bread manufacture where it aids the leavening process.

The negatively charged oxygen atoms in pyrophosphate are capable of binding a metal ion such as magnesium and calcium. The binding of such ions can have a dramatic effect on the standard free energy of hydrolysis of the phosphoanhydride. The free energy of hydrolysis of 36 kJ mol^{-1} for the hydrolysis of sodium pyrophosphate falls to 19 kJ mol^{-1} for magnesium pyrophosphate.

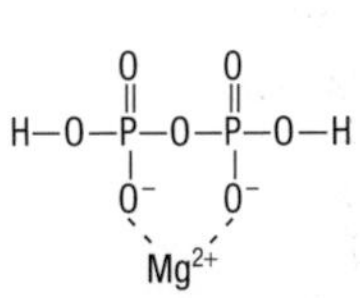

H–O–P(=O)(O⁻)–O–P(=O)(O⁻)–O–H with Mg^{2+} bound to both O⁻

Binding of magnesium strains the pyrophosphate link

Several factors make phosphoanhydride bond hydrolysis capable of yielding large amounts of free energy under standard conditions. Two of the factors will be discussed here. The phosphorus–oxygen bond is polar with both phosphorus atoms containing positive dipoles; these adjacent dipoles cause electrostatic repulsion between atoms which is lost when hydrolysis occurs. The two orthophosphates formed on hydrolysis are stabilised by internal resonance.

9.6 PHOSPHATE ESTERS

Phosphate esters can be formed by reaction between alcohols and phosphate:

$$ROH + HOPO_3^{2-} \rightleftharpoons ROPO_3^{2-} + H_2O$$

The phosphate ester has a similar structure to organic esters (see margin). The phosphate esters of the hydroxyls in carbohydrates and sugars form important intermediates in glycolysis and gluconeogenesis. For instance glucose-1′-phosphate, glucose-6′-phosphate and glyceraldehyde-3′-phosphate are all important intermediates in glycolysis. It is necessary to identify which carbon contains the phosphate ester.

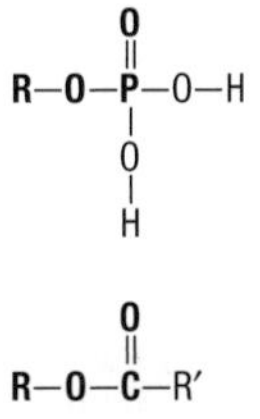

R–O–P(=O)(O–H)–O–H

R–O–C(=O)–R′

Similarity of organic and phosphate esters

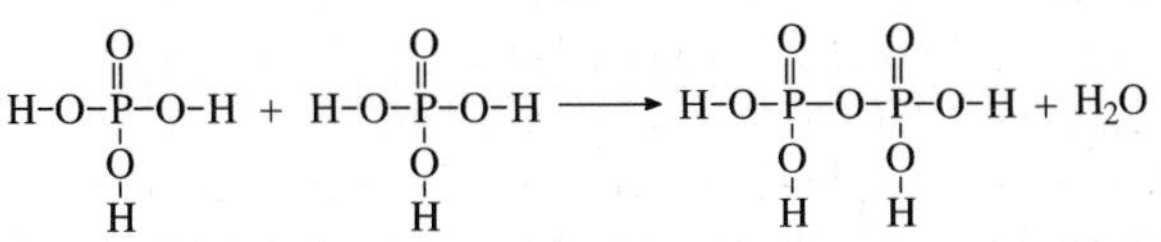

$$\text{H-O-P(=O)(OH)-O-H} + \text{H-O-P(=O)(OH)-O-H} \longrightarrow \text{H-O-P(=O)(OH)-O-P(=O)(OH)-O-H} + H_2O$$

• **Figure 9.4** Condensation of two phosphoric acids to form pyrophosphoric acid.

$$H_3PO_4 \rightleftharpoons H^+ + H_2PO_4^- \rightleftharpoons 2H^+ + HPO_4^{2-} \rightleftharpoons 3H^+ + PO_4^{3-}$$

WORKED EXAMPLE 9.3

Some phosphate esters are shown in Figure 9.5(a). Identify which carbon has the phosphate ester attached.

• **Figure 9.5** Some phosphate esters (some hydrogens omitted for clarity).

ANSWER

The first step to numbering the position of the phosphate ester is to identify the carbon that defines the numbering. In (a) (i) the functional group is the furanose oxygen (shown in bold) and carbon 1 is shown next to it in italics. The phosphate ester is bound to carbon 5. This sugar is then a 5′-phosphate. In fact, the sugar is ribose and this molecule is ribose-5′-phosphate. In compound (a) (ii) the phosphate ester number is worked out in a similar manner. The aldehyde shown in bold is at carbon 1 and the phosphate ester is therefore at position 3. This compound is D-glyceraldehyde-3′-phosphate.

QUESTION 9.4

Some phosphate esters are shown in Figure 9.5(b). Identify which carbon has the phosphate ester attached.

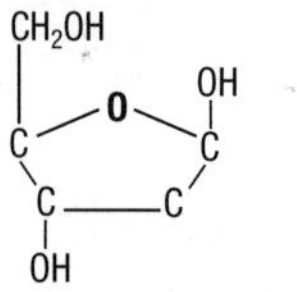

Deoxyribose: the sugar found in DNA

Sugars are often found esterified to pyrophosphate, forming sugar diphosphates, or to triphosphate, yielding sugar triphosphates. The position of such esters is identified in a similar way to the sugar phosphates.

Ribose is one of the building blocks of nucleotides which are themselves the building blocks of RNA. In nucleotides, the ribose is covalently bonded at carbon 1 to one of four relevant bases (see Figure 9.6 for ATP). In ribonucleotides the bases are the purines guanine and adenine or the pyrimidines uracil and cytosine. In the deoxynucleic acids the ribose is modified to deoxyribose (see margin) by the removal of the hydroxyl from carbon 2 and the base uracil is replaced by thymine.

Proteins are capable of forming phosphate esters by making use of the hydroxyl groups found in serine and tyrosine side chains. Phosphate ester formation can markedly affect enzyme activity in enzymes such as glycogen phosphorylase by causing a major conformational change in the protein. The phosphate ester can be esterified further to form a phosphodiester:

$$\mathrm{ROH + R'OPO_3^{2-} \rightleftharpoons RO(R'O)PO_2^- + OH^-}$$

The important phosphodiester chemical messengers cAMP and cGMP

Important phosphodiesters include the cyclic nucleotides such as cyclic adenosine-3′,5′-monophosphate (cAMP) and cyclic guanosine-3′,5′-monophosphate (cGMP) which serve as chemical messengers. Phosphodiesters are also important as the bridge between

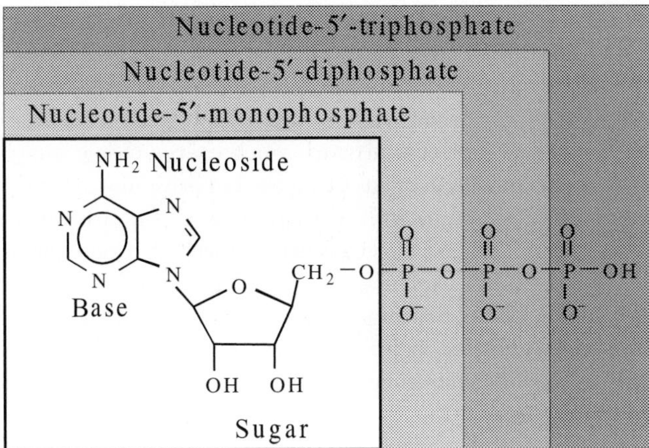

• **Figure 9.6** The relationship between base, nucleoside and nucleotide as shown for adenosine-5′-triphosphate (ATP).

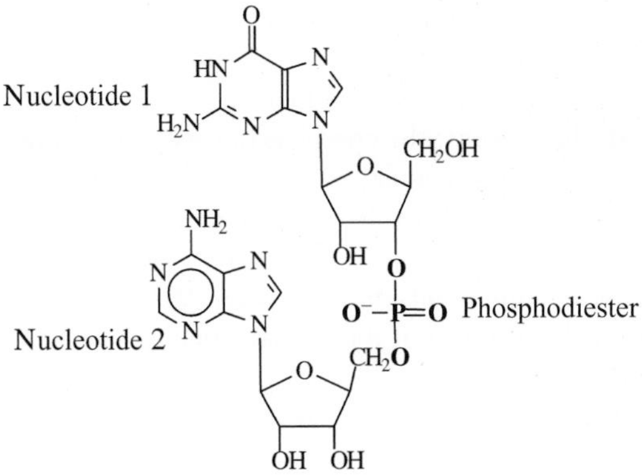

• **Figure 9.7** The phosphodiester bridge (in bold) in nucleic acids such as DNA and RNA (some hydrogens omitted).

nucleotides in DNA and RNA (Figure 9.7). The p*K* of the titratable phosphodiester oxygen is between 1 and 2. Thus only the phosphodiester ion effectively exists at physiological pHs. This gives DNA and RNA an overall negative charge at neutral pH. The phosphate groups in phosphodiesters can form salt links with positively charged binding proteins, such as the histones that bind to the DNA double helix.

Phosphodiester bridges form part of the hydrophilic part of phospholipids. Lipids are dealt with in Chapter 7. Diacylglycerols contain one free hydroxyl group that can be esterified by phosphate to form a phosphatidic acid:

$$\text{diacylglycerol} + \text{phosphate} \rightarrow \text{phosphatidic acid} + \text{water}$$

The phosphate ester oxygen can be esterified further by alcohols; for example by ethanolamine to form phospholipids such as phosphatidylethanolamine:

$$\text{phosphatidic acid} + \text{ethanolamine} \rightarrow \text{phosphatidylethanolamine} + \text{water}$$

QUESTION 9.5

(a) Name three biologically important classes of phosphodiester.

(b) Explain why phosphodiesters are negatively charged at neutral pH.

9.7 THE ROLE OF PHOSPHATE ESTERS AND ATP IN CELLULAR ENERGY METABOLISM

One of the major biological roles of phosphate esters is as intermediaries in the energy metabolism of the cell. Cells need to convert the energy from oxidation reactions into a form that can be used to drive useful work. Such energy changes might involve the energy released from the oxidation of fats being used to drive muscle contraction. Living systems are temperature sensitive, with many processes being irreversibly damaged if heat is used as the energy link (as in a steam-driven turbine). Consequently, a chemical intermediate is used to transfer the energy from oxidation to carry out useful work. The chemical link is often the phosphorylation and dephosphorylation of organic compounds. The standard free energy for the hydrolysis of phosphate bonds in some compounds is shown in Table 9.2. The most important compound used in energy transfer is adenosine-5′-triphosphate (ATP; Figure 9.6).

ATP has two properties that make it a useful biological energy vehicle. The free energy of hydrolysis means that at cellular concentrations of ATP and adenosine-5′-diphosphate (ADP) as much as 40 kJ mol^{-1} can be liberated and the reaction might be expected to occur spontaneously. ATP hydrolysis, however, occurs at only a negligible rate compared with metabolic processes in neutral solutions due to the high activation energy barrier to hydrolysis. Thus, the energy is not wasted by unwanted hydrolysis but can be liberated when needed by cellular enzymes. The second reason is that the energy of ATP hydrolysis is large enough to drive many energetically unfavourable reactions such as muscle contraction, maintenance of ion gradients across membranes, and biosynthesis reactions. Other compounds shown in Table 9.2, such as phosphoenolpyruvate, have a higher energy of hydrolysis but do not have the high activation energy shown by ATP and are less stable.

The actual energy that can be obtained by ATP hydrolysis is complex and depends on a number of factors. One factor is that most cells maintain the concentration ratio of ATP:ADP well above the standard conditions. This increases the energy that can be obtained from the hydrolysis reaction (see Chapter 11). Another reason is that the ATP phosphates can bind metal ions, particularly magnesium, and this affects the energy of hydrolysis (see section 9.5). A third factor is that both reactants and products are acids and thus the reaction will be affected by the pH of the solution (depending on the pH, the hydrogen ion, H^+, is either a product or a reactant).

Table 9.2 Free energy of hydrolysis of some organic phosphates (pH 7.0 excess magnesium)

Hydrolysis reaction (water not included)		*$-\Delta G°$(kJ mol^{-1})*
Reactant	*Products*	
Phosphoenolpyruvate	Pyruvate + phosphate	62
Creatine phosphate	Creatine + phosphate	43
Adenosine-5′-triphosphate	Adenosine-5′-diphosphate + phosphate	32
Adenosine-5′-triphosphate	Adenosine-5′-monophosphate + pyrophosphate	32
Glucose-1′-phosphate	Glucose + phosphate	21
α-Glycerophosphate	Glycerol + phosphate	10
Glucose-6′-phosphate	Glucose + phosphate	13

■ SUMMARY

Phosphorus and sulphur both have multiple valencies resulting from use of the d orbitals in the third valence shell. Assimilated sulphur is normally valency six and must be reduced before biosynthesis of sulphur-containing compounds. Thiols are involved in reduction–oxidation control within cells as well as being structural components of proteins. Thiols are capable of esterification to form thioesters, and thioesters provide alternative routes for cellular metabolism. Sulphate esters are important structural components of peptidoglycans providing a site for interaction with charged species. Phosphoric acid is trivalent. Two of the protons from phosphoric acid are dissociated at pH 7.4 to give the hydrogenphosphate group. Phosphate salts are important structural components of bones. The structure of the phosphate group is stabilised by delocalisation of electrons. Two or more phosphates can be condensed together to form pyrophosphate, triphosphate and polyphosphates. The phosphoanhydride bridge in pyrophosphate is unstable and hydrolysis can provide the energy for work. Phosphate is capable of reacting with hydroxyls to form phosphate esters. Phosphate esters are important intermediates in cellular metabolism. Enzyme phosphorylation is used to regulate metabolic pathways. Sugar phosphates are components of nucleotides and the phosphate is referenced to the carbon number of the esterified hydroxyl. The nucleotide ATP carries the energy generated by oxidation reactions to do useful work. Cyclic nucleotides can act as cellular messengers, controlling the esterification of proteins that carry out metabolic reactions. Nucleotides can be joined together by phosphodiester bridges to form RNA and DNA. Phosphodiesters form part of phospholipids that are commonly found in biological membranes.

■ SUGGESTED FURTHER READING

Corbridge, D.E.C. (1995) *Phosphorus – an Outline of its Chemistry, Biochemistry and Technology*, 5th edn. Elsevier, Amsterdam.

Fox, M.A. and Whitesell, J.K. (1997) *Organic Chemistry*, 2nd edn., pp. 131–133, 622–623. Jones and Bartlett.

■ END OF CHAPTER QUESTIONS

Question 9.6
(a) Draw the structure of reduced and oxidised glutathione.
(b) Name the sulphur-containing functional groups in reduced and oxidised glutathione.

Question 9.7
Explain why phosphate is an excellent buffer at pH 7.2. You may need to refer to Chapter 5 to illustrate your answer.

Question 9.8
(a) Draw pyrophosphoric acid and label the phosphoanhydride bond.
(b) List two factors that make phosphoanhydride bond hydrolysis capable of yielding large amounts of free energy.
(c) State three factors that can affect the free energy of hydrolysis of phosphoanhydrides.

Question 9.9
(a) Draw a general reaction for the formation of phosphodiesters from a phosphate ester.
(b) Give three functions of phosphodiesters.

10 OXIDATION AND REDUCTION REACTIONS

■ 10.1 INTRODUCTION

Many of the chemical reactions carried out in living cells involve oxidation and reduction. Oxidation reactions are often involved with the release of energy from energy storage compounds such as carbohydrates. It is important for us to be able to calculate how much energy oxidation reactions can yield. In a similar way some reduction reactions in biology, such as the dark reactions of photosynthesis, need energy to drive them. The oxidation and reduction processes in biology are largely reversible reactions. As scientists we like to be able to predict in which direction such reversible reactions are going to proceed and what effect changes in concentration have upon the direction of these reactions.

■ 10.2 OXIDATION IS LINKED TO REDUCTION

Let us look at the complete oxidation of the sugar glucose:

$$\underset{\text{glucose}}{C_6H_{12}O_6} + \underset{\text{oxygen}}{6O_2} \rightleftharpoons \underset{\text{carbon dioxide}}{6CO_2} + \underset{\text{water}}{6H_2O}$$

The carbon and hydrogen atoms in glucose have been **oxidised** by bonding to oxygen. The oxygen atoms have been **reduced**. The two processes are interlinked. An oxidation of one molecule in a chemical reaction *must* be linked to a reduction of another molecule (or another part of the same molecule). In this example glucose is oxidised whilst oxygen is reduced. The process is often called a reduction–oxidation process to emphasise this link. Reduction–oxidation is abbreviated to **redox** and chemical reactions involving reduction–oxidation are known as redox reactions. We can write this reaction as a general process involving the reduction of molecule A and oxidation of molecule B:

reduction–oxidation ↘ ↙ redox

$$A_{Ox} + B_{Red} \rightleftharpoons B_{Ox} + A_{Red}$$

Molecule A is acting to oxidise molecule B and is called an oxidising agent or **oxidant**. Molecule B reduces molecule A and is therefore called a reducing agent or **reductant**. A molecule acting as an oxidant must itself become reduced.

Some oxidation reactions are unwanted and can result in rancidity of fats and increased risks of cancer. Protection from the consequences of such reactions can be obtained by using molecules that are more easily oxidised than the chemicals in the unwanted reaction. Such molecules are called **antioxidants** and include ascorbic acid (vitamin C) and tocopherol (vitamin E).

10.3 THE CHEMICAL CHANGES IN THE REDOX PROCESS

The complete oxidation of glucose is just one example of a redox reaction. Some general rules about redox reactions can be obtained by thinking about the changes in the structure of the chemicals involved.

In glucose many of the carbon atoms are bonded to hydrogen whereas after oxidation each carbon and hydrogen atom is completely bound to oxygen. Oxidation can be seen as the removal of either electrons or hydrogen or as the addition of oxygen to a molecule. The opposite is true for reduction. We can tell when a compound has been oxidised by looking for these signs. We must be careful as not all hydrogen additions are reduction reactions. In accepting a hydrogen *ion*, a base (see section 5.4) does not become reduced. The best evidence that a chemical has been reduced is the addition of an electron.

Oxidation:
removing electrons
adding oxygen
removing hydrogen
Reduction:
adding electrons
adding hydrogen
removing oxygen

WORKED EXAMPLE 10.1

(a) In the following reactions decide which molecule has been reduced, and which oxidised.
   (i) malate + NAD^+ → oxaloacetate + NADH + H^+
   (ii) cytochrome b (Fe^{2+}) + cytochrome c_1 (Fe^{3+}) → cytochrome b (Fe^{3+}) + cytochrome c_1 (Fe^{2+})
(b) In reactions (i) and (ii) which reactant is the oxidant?

ANSWER

(i) The NAD^+ has had a proton and two electrons added and must be reduced. As each reduction is linked to an oxidation then malate must have been oxidised in conversion to oxaloacetate.
(ii) The cytochrome c_1 has had an electron added and must have been reduced. The cytochrome b has had an electron removed and must have been oxidised.
   (b) The oxidant is reduced (or oxidises the other compound) and so the oxidant is NAD^+ in (i) and cytochrome c_1 in (ii).

QUESTION 10.1

(a) In the following reactions decide which molecule has been reduced, and which oxidised.
   (i) ethanal + NAD^+ → ethanoate + NADH
   (ii) cytochrome c (Fe^{2+}) + Cu^{2+} → cytochrome c (Fe^{3+}) + Cu^+
(b) In reactions (i) and (ii) which reactant is the reductant?

10.4 SPLITTING REDOX REACTIONS

In section 10.2 it was easier to see which component had been oxidised and which reduced in a redox reaction by considering the fate of each molecule separately. Let us look at the oxidation of malate in the Krebs (TCA) cycle again:

$$\text{malate} + NAD^+ \rightarrow \text{oxaloacetate} + NADH$$

This reaction can be split by looking at malate and NAD^+ separately:

$$\begin{array}{c} COOH \\ | \\ \mathbf{HOCH} \\ | \\ CH_2 \\ | \\ COOH \\ \text{malate} \end{array} \rightleftharpoons \begin{array}{c} COOH \\ | \\ O{=}C \\ | \\ CH_2 \\ | \\ COOH \\ \text{oxaloacetate} \end{array} + \mathbf{2H^+} + 2e^-$$

$$NAD^+ + 2e^- + 2H^+ \rightleftharpoons NADH + H^+$$

Here the malate has lost two electrons and two protons (shown in bold) and is clearly oxidised whilst the NAD^+ has gained two electrons and a proton and is reduced.

The oxidation of malate or the reduction of NAD^+ are examples of **half-reactions**. A general formula can be written for a half-reaction which is:

$$Ox + ne^- + mH^+ \rightarrow Red$$

where n is the number of electrons and m the number of protons involved in the reaction. It is conventional to write all half-reactions in this way.

WORKED EXAMPLE 10.2

Break each of the following reactions into half-reactions and represent each half-reaction using the general formula.

(i) $HOOC{-}\mathbf{CH_2}{-}\mathbf{CH_2}{-}COOH + FAD \rightarrow HOOC{-}CH{=}CH{-}COOH + FADH_2$
succinate — fumarate

(ii) cytochrome b (Fe^{2+}) + cytochrome c_1 (Fe^{3+}) → cytochrome b (Fe^{3+}) + cytochrome c_1 (Fe^{2+})

ANSWER

(i) The key to the number of electrons transferred here is to remember that each covalent bond contains *two* electrons. In succinate the two central carbons (shown in bold) are joined by a single bond, but each is also bonded to two hydrogen atoms. The central carbons in fumarate contain a double carbon bond with each carbon bonded to only one hydrogen. Fumarate has gained one covalent bond between the central carbons but lost two carbon–hydrogen bonds and therefore has one covalent bond less. This means that two electrons have been lost from the molecule. Two protons have also been lost. The half-reaction is therefore:

$$\text{succinate} \rightarrow \text{fumarate} + 2e^- + 2H^+$$

The convention is to write the half-reaction as a reduction and so we must rewrite this as:

$$\text{fumarate} + 2e^- + 2H^+ \rightarrow \text{succinate}$$

The FAD half-reaction is: $FAD + 2e^- + 2H^+ \rightarrow FADH_2$

(ii) cytochrome c_1 (Fe^{3+}) + e^- → cytochrome c_1 (Fe^{2+})
cytochrome b (Fe^{3+}) + e^- → cytochrome b (Fe^{2+})

QUESTION 10.2

Break each of the following reactions into half-reactions and represent each half-reaction using the format for a general formula.

(i) malate + NAD^+ → oxaloacetate + NADH

(ii) cytochrome c (Fe^{2+}) + Cu^{2+} → cytochrome c (Fe^{3+}) + Cu^+

10.5 STANDARDISING REDOX HALF-REACTIONS

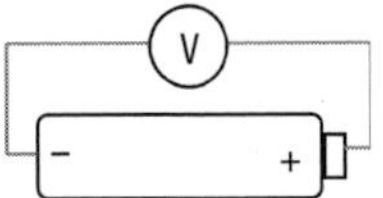

A battery with potential difference measured in volts using a voltmeter

A redox reaction involves the movement of electrons from one molecule to another. The movement of electrons is an electrical current (measured in amps) and a pair of redox half-reactions acts very much like a battery. In fact all conventional batteries involve redox reactions in their construction. The potential difference between the poles of a battery is measured in volts (V) using a voltmeter and the potential difference between two redox half-reactions is also measured in volts.

All redox half-reactions are written in the same direction to allow comparison between them. It is possible to measure the direction of electron flow of any half-reaction by comparing it with a standard half-reaction. The standard half-reaction that is used is:

$$2H^+ + 2e^- \rightarrow H_2$$

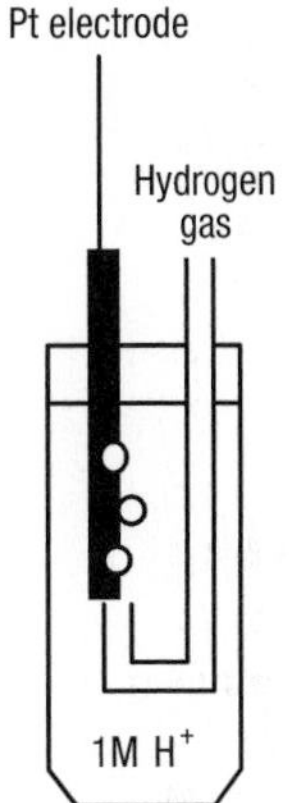

A hydrogen electrode

This is the hydrogen/proton half-reaction and can be set up as an electrode. The hydrogen electrode is a solution of strong acid at 1 mol dm^{-3} through which hydrogen gas is bubbled at atmospheric pressure. The concentrations of the oxidised and reduced forms are standardised at 1 mol dm^{-3} for solutes and atmospheric pressure for gases. The electrical connection is made using an inert substance such as platinum. This electrode is abbreviated to Pt | H_2(g) | H^+. The vertical lines indicate junctions between the phases. Redox half-reactions will occur in the same phase and a comma is used to separate the reduced and oxidised forms. The potential of the hydrogen electrode is given a value of 0.0 V. The value 1 mol dm^{-3} [H^+] is not biologically relevant and so conversion to pH 7.0 is normally made. At pH 7.0 the potential of the hydrogen electrode converts to −0.42 V. This value is a standard value and is called a **standard reduction potential**. The symbol for the standard reduction potential at pH 7.0 is E°.

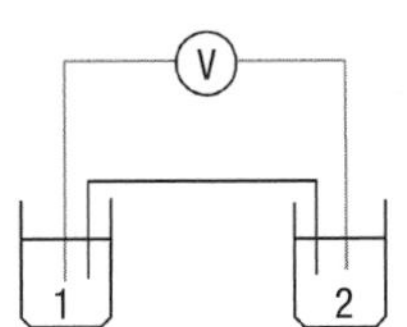

Two redox half-reactions have their potential difference measured in volts using a voltmeter

Other electrodes containing half-reactions can be compared with this standard electrode by making up a circuit containing both half-reactions. The standard reduction potential can then be measured with a voltmeter. The standard reduction potentials for some biologically relevant half-reactions are shown in Table 10.1.

10.6 PREDICTING ELECTRON FLOW

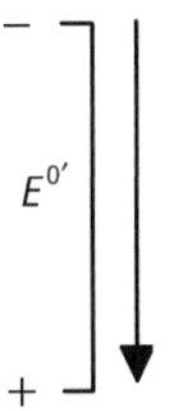

Exergonic direction of electron flow

It is important to be able to predict in which direction electrons will flow in a redox reaction. The standard reduction potential for the two half-reactions can be used to do this. Electron flow will be from less positive E° to more positive values (it may help to think of an electron bearing a negative charge and therefore moving towards the more positive half-reaction).

Table 10.1 Standard reduction potentials of some half-reactions

Oxidant	*Reductant*	*Number of electrons (n)*	$E^{o\prime}$ *(V)*
Succinate + CO_2	α-Ketoglutarate	2	−0.67
Acetate	Acetaldehyde	2	−0.60
Ferredoxin (oxidised)	Ferredoxin (reduced)	1	−0.43
2 H^+	H_2	2	−0.42
NAD^+	NADH + H^+	2	−0.32
$NADP^+$	NADPH + H^+	2	−0.32
Lipoate (oxidised)	Lipoate (reduced)	2	−0.29
Glutathione (oxidised)	Glutathione (reduced)	2	−0.23
FAD	$FADH_2$	2	−0.22
Ethanal	Ethanol	2	−0.20
Pyruvate	Lactate	2	−0.19
Oxaloacetate	Malate	2	−0.17
Fumarate	Succinate	2	0.03
Cytochrome b (+3)	Cytochrome b (+2)	1	0.07
Dehydroascorbate	Ascorbate	2	0.08
Ubiquinone (oxidised)	Ubiquinone (reduced)	2	0.10
Cytochrome c (+3)	Cytochrome c (+2)	1	0.22
Cu (+2)	Cu (+1)	1	0.34
Fe (+3)	Fe (+2)	1	0.77
$\frac{1}{2}O_2/H_2O$	H_2O	2	0.82

WORKED EXAMPLE 10.3

(a) Predict which way electrons will flow between the following redox half-reactions:
   (i) malate, oxaloacetate and NADH, NAD^+
   (ii) ubiquinone$_{(red)}$, ubiquinone$_{(ox)}$ and cytochrome c$_{(red)}$, cytochrome c$_{(ox)}$

(b) Which half-reaction will become oxidised and which reduced for (i) and (ii)?

ANSWER

(a) Look up values of E° in Table 10.1 for the redox half-reactions.
   (i) This is −0.17 V for malate, oxaloacetate and −0.32 V for NADH, NAD^+. Electrons will flow from NADH (less positive) to oxaloacetate (more positive).
   (ii) This is +0.11 V for ubiquinone$_{(Red)}$, ubiquinone$_{(Ox)}$ and +0.22 V for cytochrome c$_{(Red)}$, cytochrome c$_{(Ox)}$. Electrons will flow from ubiquinone to cytochrome c.

(b) (i) NADH will become oxidised and oxaloacetate reduced.
   (ii) Ubiquinone will become oxidised and cytochrome c reduced.

QUESTION 10.3

(a) Predict which way electrons will flow between the following redox half-reactions:

(i) lactate, pyruvate and NADH, NAD^+

(ii) $ascorbate_{(Red)}$, $dehydroascorbate_{(Ox)}$ and Fe^{2+}, Fe^{3+}

(b) Which half-reaction will become oxidised and which reduced for (i) and (ii)?

10.7 FREE ENERGY AND STANDARD REDUCTION POTENTIALS

The free energy change ΔG° can be predicted from the difference in reduction potentials between half-reactions. This is given by the equation (see Appendix, Derivation 10.1):

$$\Delta G^\circ = -n\Delta E^\circ F$$

where n is the number of electrons transferred, ΔE° is the difference in standard reduction potential between the two half-reactions, measured in volts, and F is a constant called the Faraday constant. The Faraday constant can be viewed as a conversion constant for volts into joules and has a value of 96 500 J V^{-1} mol^{-1}. ΔE° is always calculated by subtracting donor E° from acceptor E°. It must be remembered that a negative value of ΔG, which indicates that a reaction can occur spontaneously, is obtained from a positive value of ΔE.

WORKED EXAMPLE 10.4

Calculate the ΔG° for the following redox reactions:

(i) NAD^+ + malate + H^+ → oxaloacetate + NADH

(ii) succinate + FAD → fumarate + $FADH_2$

ANSWER

Look up values of E° and the number of electrons transferred in Table 10.1 for the redox half-reactions.

(i) This is −0.17 V for malate, oxaloacetate and −0.32 V for NADH, NAD^+. In the direction that the reaction is written, two electrons will flow from malate to NAD^+ and so malate is the donor. ΔE is therefore −0.32 − (−0.17) = −0.15 V.

Substitution into $\Delta G^\circ = -n\Delta EF$ gives

$\Delta G^\circ = -2 \times -0.15 \times 96\,500$ J mol^{-1} = +28 950 J mol^{-1}

(ii) This is 0.03 V for fumarate, succinate and −0.22 V for FADH, $FADH_2$. In the direction that the reaction is written, two electrons will flow from succinate to FAD and so succinate is the donor. ΔE is therefore −0.22 − (0.03) = −0.25 V.

Substitution into $\Delta G^\circ = -n\Delta EF$ gives

$\Delta G^\circ = -2 \times -0.25 \times 96\,500$ J mol^{-1} = +48 250 J mol^{-1}

The positive value of ΔG shows that these reactions cannot occur spontaneously.

QUESTION 10.4

Calculate the ΔG for the following redox reactions:

(i) pyruvate + NADH → lactate + NAD^+

(ii) Cu^{2+} + Fe^{2+} → Fe^{3+} + Cu^+

10.8 REDOX REACTIONS AND NON-STANDARD CONDITIONS

During the course of redox reactions the concentrations of reactants are changing and standard reduction potentials must therefore be calculated by imposing a potential exactly opposite to that created by the two half-reactions. This stops electrons moving and keeps the concentrations at their standard value. It is also possible to take into account non-standard conditions when calculating the reduction potential by using the **Nernst equation** for each of the half-reactions. The Nernst equation is (see Appendix, Derivation 10.2):

$$\Delta E = \Delta E^\circ + \frac{2.3RT}{nF}\log_{10}\frac{[\mathrm{Ox}]}{[\mathrm{Red}]}$$

where [Ox] and [Red] are the concentrations (mol dm^{-3}) of oxidised and reduced species, R is the gas constant 8.3 J K^{-1} mol^{-1} and T is the temperature in kelvin. The body temperature of 310 K gives a value for 2.3 RT/F of 0.06 V and so the equation simplifies to:

$$\Delta E = \Delta E^\circ + \frac{0.06}{n}\log_{10}\frac{[\mathrm{Ox}]}{[\mathrm{Red}]}$$

It should be noted that if [Ox] = [Red] then $\Delta E = \Delta E^\circ$.

WORKED EXAMPLE 10.5

Calculate the ΔG for the following redox reaction:

$$\mathrm{NAD^+ + malate + H^+ \rightarrow oxaloacetate + NADH}$$

when [NAD^+] is 10^{-3} mol dm^{-3} and [NADH] is 10^{-5} mol dm^{-3} at 37°C.

ANSWER

Look up values of E° in Table 10.1 for the redox half-reactions.
This is –0.17 V for malate, oxaloacetate and –0.32 V for NADH, NAD^+.
Use the Nernst equation to recalculate ΔE for the NADH/NAD^+ half-reaction to take into account the change in concentrations of NADH and NAD^+.

$$\Delta E = -0.32 + \frac{0.06}{2}\log_{10}\frac{10^{-3}}{10^{-5}}$$

$$= -0.32 + 0.03(+2)\ \mathrm{V}$$

$$= -0.26\ \mathrm{V}$$

In the direction that the reaction is written, two electrons will flow from malate to NAD^+ and so malate is the donor. ΔE is therefore –0.26 – (–0.17) = –0.07 V.
Substitution into $\Delta G = -n\Delta EF$ gives
$\Delta G = -2 \times -0.07 \times 96\,500$ J mol^{-1} = +13 510 J mol^{-1}

The worked example 10.5 is of interest as in cells the equilibrium position of endergonic reactions can be changed by altering the concentration of reactants and products. In this case the endergonic TCA cycle reaction of malate to oxaloacetate can be made more favourable if the NADH concentration is kept low by its constant oxidation in the mitochondrion.

QUESTION 10.5

Calculate the ΔG for the following redox reaction:

$$\text{pyruvate} + \text{NADH} \rightarrow \text{lactate} + \text{NAD}^+$$

when [pyruvate] is 3×10^{-3} mol dm^{-3} and [lactate] is 6×10^{-4} mol dm^{-3} at 37°C.

■ SUMMARY

Reduction and oxidation reactions are important in biology. Oxidation and reduction are the loss or gain of electrons respectively by a molecule. Reduction of one compound is always linked to the oxidation of another compound. Redox reactions can be broken down into half-reactions by considering each compound separately. The direction of electron flow between half-reactions can be predicted from standard reduction potentials, and can be adjusted for non-standard conditions using the Nernst equation.

■ SUGGESTED FURTHER READING

Many of the thermodynamic principles underpinning this chapter are well covered in the following texts:

Wrigglesworth, J. (1997) *Energy and Life*, pp. 47–56. Taylor and Francis, London.
Atkins, P.W. (1994) *The Elements of Physical Chemistry*, Ch. 6, 5th edn. Oxford University Press.
Price, N.C. and Dwek, R. (1979) *Physical Chemistry for Biochemists*, Ch. 8, 2nd edn. Oxford University Press.

■ END OF CHAPTER QUESTIONS

Question 10.6 Glutathione reductase catalyses electron transfer between glutathione and NADPH. The reaction can be summarised as:

$$\text{glutathione}_{(Ox)} + \text{NADPH} + \text{H}^+ \rightarrow 2\ \text{glutathione}_{(Red)} + \text{NADP}^+$$

(a) Under standard conditions in which direction will the reaction proceed?
(b) Calculate the free energy change available under standard conditions.
(c) The ratio in living cells of reduced to oxidised glutathione is normally 500:1. Calculate the free energy change for the reaction at this ratio of [Red]:[Ox].

Question 10.7 In mitochondrial electron transport, electrons are transferred between ubiquinone and cytochrome c.

(a) In which direction would electron flow be energetically favourable?
(b) Calculate the $\Delta G°$ under standard conditions for electron flow from ubiquinone to cytochrome c.

11 ENERGY

■ 11.1 INTRODUCTION

Living things require energy for a variety of purposes. Plants require energy from sunlight in order to photosynthesise. Animals and plants both require energy to grow, reproduce, repair and maintain their tissues. The energy for these latter processes is derived from food by chemical reactions. In this chapter, we examine the role of energy in chemical change, a subject technically known as **chemical thermodynamics**.

■ 11.2 THE FIRST LAW OF THERMODYNAMICS

Some types of energy

Kinetic energy
Gravitational (potential) energy
Heat energy
Chemical energy
Electrical energy
Magnetic energy
Light energy
Sound energy
Nuclear energy

The first law of thermodynamics is also known as the energy conservation law. It states:

> Energy cannot be created or destroyed, but only transformed from one kind of energy to another.

For example, a muscle converts chemical energy, stored in the chemical ATP, into kinetic energy, the energy associated with movement. Such a conversion of energy is called an **energy transduction**, and the apparatus that brings about the transduction is called an **energy transducer**.

WORKED EXAMPLE 11.1

(a) What sort of energy transducer converts sound energy into electrical energy in a vertebrate animal?

(b) In the overall synthesis of sugars by photosynthesis what energy transduction has occurred?

ANSWER

(a) An ear converts sound energy into electrical energy in the nerve impulses sent to the brain.

(b) During photosynthesis, plants convert the energy in sunlight into chemical energy stored in the chemical bonds of glucose and starch.

QUESTION 11.1

(a) What sort of device converts electrical energy into light energy?

(b) What energy conversion occurs during exercise?

■ 11.3 UNITS OF ENERGY

In the SI system, energy is measured in **joules** (symbol J). Many chemical reactions produce hundreds of thousands of joules of energy, so it is often convenient to use the larger unit, the kilojoule, kJ, equal to 1000 J.

In order to compare the energy changes brought about by different reactions on a consistent basis, we also need to take into account the quantity of reactants used. Energy changes in chemical reactions are therefore usually measured in kilojoules per mole of reactants, kJ mol^{-1}.

■ 11.4 MEASUREMENT OF ENERGY

It is important to realise that we can never measure the *total* energy contained in a given substance. The total energy will consist of chemical energy stored in the chemical bonds present in the material, kinetic energy due to the motions of the atoms and electrons, electrical energy due to the charges on the electrons and protons, nuclear energy due to the interactions between protons and neutrons in the nuclei of the atoms, and so on. Instead of measuring the total energies present, we must content ourselves with measuring energy *changes*.

■ 11.5 INTERNAL ENERGY, *U*, AND ENTHALPY, *H*

Even though we cannot measure the total energy present in a substance, we still have a name for this energy: **internal energy**, given the symbol U. We can only measure changes in internal energy, and these changes are given the symbol ΔU, pronounced 'delta U'.

All chemical changes lead to changes in internal energy. In most cases, this energy appears in two forms: as heat and as work. Work is the energy which must be expended to *do* something. In the biochemical context, this work may consist of having to push against atmospheric pressure so that gases produced in a reaction can escape.

Because most chemical and biochemical reactions occur at atmospheric pressure and have to do work against this pressure, chemists and biochemists usually use an alternative to internal energy as a measure of the energy available from a reaction. This measure is called **enthalpy**, and is given the symbol H. Enthalpy changes, which are all we can measure, are given the symbol ΔH. The enthalpy of a reaction is the amount of energy left after the work of pushing against atmospheric pressure has been done. It should be noted that for reactions involving only solids and liquids, in which volume changes are usually very small, ΔU and ΔH will be almost identical.

Reactions which release heat as they occur, so that the reaction mixture gets warmer, are called **exothermic**, and the enthalpy change is given a **negative** sign to indicate that the reactants have lost energy in being transformed into products. Reactions which take in heat are called **endothermic**. Such reactions cause the reaction mixture to cool and the enthalpy change is given a positive sign to indicate that the reactants have gained energy in being transformed into products.

■ 11.6 CALORIMETRY

The energy changes involved in chemical reactions can be investigated using the methods of calorimetry. In these methods, a chemical reaction is carried out in a special container called a **calorimeter**.

The heat produced by the reaction is transferred to a known mass of water and the temperature rise of the water is observed. Knowing the mass of water, its heat capacity and the temperature rise, the amount of heat given out by the reaction can be calculated.

The heat capacity of water is the amount of heat which must be supplied to 1 g of water to raise its temperature by 1 K.

> Heat gained by water and calorimeter = Heat released by reaction = Change in temperature (ΔT) × Mass of water in grams (m) × Heat capacity of water in joules per kelvin per gram (C).

If the reaction is carried out in a sealed container (called a *bomb calorimeter*) the measured quantity of heat will be ΔU for the reaction, whereas if the calorimeter is open to the atmosphere, the measured value will be ΔH.

WORKED EXAMPLE 11.2

2.3 g of ethanol are burned in an open calorimeter. The heat produced raises the temperature of 1000 g of water by 16.3 K. Calculate the enthalpy of combustion of ethanol in kJ mol^{-1}, given that the heat capacity of water = 4.18 J K^{-1} g^{-1}.

ANSWER

Heat taken up by the water = 16.3 × 1000 × 4.18 = 68 134 J

This was released from 2.3 g of ethanol

Molar mass of ethanol = 46 g mol^{-1}

$$2.3 \text{ g of ethanol} = \frac{2.3}{46} \text{ moles} = 0.05 \text{ moles}$$

$$\text{Energy produced per mole of ethanol} = \frac{68\,134}{0.05} \text{ J mol}^{-1} = 1\,362\,680 \text{ J mol}^{-1}$$

Converting to kJ mol^{-1} and rounding off to a sensible number of digits gives:

Enthalpy of combustion of ethanol = 1363 kJ mol^{-1}

QUESTION 11.2

When 1.8 g of glucose are burned in an open calorimeter, the temperature rise of 500 g of water is found to be 13.5 K. Calculate the enthalpy of combustion of glucose in kJ mol^{-1}.

■ 11.7 HESS'S LAW

Energy

A

C

B

Hess's law states:

> The energy change involved in going from one state to another is independent of the route taken between the two states.

Consider the diagram on the left. State A is at higher energy than state B, and state C is at some intermediate energy. If we follow the arrows round from A through B and C back to A, the overall energy change must be zero, since we are back where we started from. Hess's law tells us that the energy released in going from A to B is the same as the energy required to go from B to A via C. If this were not the case, then it would be possible to continuously extract energy from the system. For example, if more energy was given out in going from A to B than was required to go from B to A via C, then every time we went round the cycle, we would find that we had some energy left over.

This would violate the first law of thermodynamics, since we would be able to create energy.

The main use of Hess's law is to enable us to calculate the energies involved in chemical reactions for which direct measurement is difficult or impossible.

WORKED EXAMPLE 11.3

The conversion of glucose, $C_6H_{12}O_6$, to pyruvic acid, $C_3H_4O_3$, is an important step in the TCA cycle:

$$C_6H_{12}O_{6(s)} + O_{2(g)} \rightarrow 2C_3H_4O_{3(l)} + 2H_2O_{(l)}$$

$$\text{glucose + oxygen} \rightarrow \text{pyruvic acid + water}$$

Calculate the enthalpy involved in this conversion, given the following enthalpies of combustion:

(1) $C_6H_{12}O_{6(s)} + 6O_{2(g)} \rightarrow 6CO_{2(g)} + 6H_2O_{(l)}$
$\Delta H_1 = -2821\ \text{kJ mol}^{-1}$

(2) $C_3H_4O_{3(l)} + \frac{5}{2}O_{2(g)} \rightarrow 3CO_{2(g)} + 2H_2O_{(l)}$
$\Delta H_2 = -1170\ \text{kJ mol}^{-1}$

ANSWER

The equation representing the conversion of glucose to pyruvic acid is:

$$C_6H_{12}O_{6(s)} + O_{2(g)} \rightarrow 2C_3H_4O_{3(l)} + 2H_2O_{(l)} \qquad \Delta H_t = ?$$

This is the target equation. The aim is to manipulate the equations (1) and (2) to produce the target equation.

The first step is to get the reactants and products of the target equation onto the right sides of the given equations. We see that the product of the target equation, pyruvic acid, is on the left-hand side of equation (2). Therefore, we must reverse this equation to get the pyruvic acid on the right-hand side:

(3) $3CO_{2(g)} + 2H_2O_{(l)} \rightarrow C_3H_4O_{3(l)} + \frac{5}{2}O_{2(g)}$
$\Delta H_3 = +1170\ \text{kJ mol}^{-1}$

Note that we must reverse the sign of ΔH when we reverse the equation.

Now all reactants and products in the target equation are on the correct sides of the given equations.

The second step is to ensure that the target numbers of reactant and product molecules are present in the given equations. The target equation contains two pyruvic acid molecules, while equation (3) contains only one.

This equation therefore must be multiplied by two:

(4) $6CO_{2(g)} + 4H_2O_{(l)} \rightarrow 2C_3H_4O_{3(l)} + 5O_{2(g)}$
$\Delta H_4 = +2340\ \text{kJ mol}^{-1}$

Note that we must multiply ΔH by two when we multiply the equation by two.

Now we have all the target reactants on the left-hand sides of their equations and all the target products on the right-hand sides. We also have them present in the correct numbers. We don't need to worry about all those species that don't appear in the target equation – they will automatically cancel out.

All we need to do now is add together all those equations that have the correct positions and numbers of reactants and products. In this case, those equations are numbers 1 and 4. We now add equation (1) to equation (4):

$$6CO_{2(g)} + 4H_2O_{(l)} + C_6H_{12}O_{6(s)} + 6O_{2(g)} \rightarrow 2C_3H_4O_{3(l)} + 5O_{2(g)} + 6CO_{2(g)} + 6H_2O_{(l)}$$

Cancelling out those species which occur on both sides of the equation leaves us with the target equation:

(5) $C_6H_{12}O_{6(s)} + O_{2(g)} \rightarrow 2C_3H_4O_{3(l)} + 2H_2O_{(l)}$

$\Delta H_t = \Delta H_1 + \Delta H_4$

$\Delta H_t = -2821 + 2340 = -481 \text{ kJ mol}^{-1}$

Thus the enthalpy of conversion of glucose to pyruvic acid is -481 kJ mol^{-1}.
The negative sign confirms that the reaction is exothermic.
Note that we always do the same thing to the ΔH values that we do to the chemical equations.

QUESTION 11.3

Anaerobic respiration in yeast involves the organism extracting energy from glucose by breaking down the molecule to form ethanol and carbon dioxide:

$$C_6H_{12}O_{6(s)} \rightarrow 2C_2H_5OH_{(l)} + 2CO_{2(g)}$$

glucose → ethanol + carbon dioxide

Calculate the energy that can be extracted by this process from one mole of glucose given the following enthalpies of combustion:

(1) $C_6H_{12}O_{6(s)} + 6O_{2(g)} \rightarrow 6CO_{2(g)} + 6H_2O_{(l)}$

$\Delta H_1 = -2821 \text{ kJ mol}^{-1}$

(2) $C_2H_5OH_{(l)} + 3O_{2(g)} \rightarrow 2CO_{2(g)} + 3H_2O_{(l)}$

$\Delta H_2 = -1368 \text{ kJ mol}^{-1}$

■ 11.8 ENTHALPIES OF FORMATION

In the examples given above, enthalpies of combustion were used to provide the data required for Hess's law calculations. Such enthalpies are readily determined and many values are listed in tables. Also frequently listed in tables of data are enthalpies of formation and these too can be used in Hess's law calculations.

The enthalpy of formation, ΔH_f, of a substance is the enthalpy change involved when one mole of the substance is made *from its elements*, the substance and all the elements from which it is made being in their standard states under the conditions of temperature and pressure employed. It follows from this definition that the enthalpy of formation of any element in its standard state is zero.

The standard state of a substance is the form which is most stable under ambient conditions. Thus, the standard state of oxygen at atmospheric pressure and 298 K is a gas. Under the same conditions, the standard state of water is a liquid and the standard state of carbon is a solid, graphite.

Using enthalpies of formation, any chemical reaction can be viewed as taking place in two stages:

1. The reactants are broken up into their elements in their standard states. This is the reverse of the process used to determine enthalpies of formation, and so the enthalpy involved in this process will be the same magnitude as that involved in the formation process but will be of reversed sign, i.e. if the enthalpy involved in this part of the process is written as ΔH_1, then

$$\Delta H_1 = -\Sigma\, \Delta H_f(\text{reactants})$$

In this equation, the symbol Σ means 'the sum of' – the enthalpies of formation of all the reactants are added together.

2. The elements in their standard states are then reassembled to give the products. The enthalpy involved in this, ΔH_2, is the enthalpy of formation of all the products.

$$\Delta H_2 = \Sigma\, \Delta H_f(\text{products})$$

The overall enthalpy of the process is the sum of the enthalpies for the two stages:

$$\begin{aligned}\Delta H_{reaction} &= -\Sigma\, \Delta H_f(\text{reactants}) + \Sigma\, \Delta H_f(\text{products})\\ &= \Sigma\, \Delta H_f(\text{products}) - \Sigma\, \Delta H_f(\text{reactants})\end{aligned}$$

WORKED EXAMPLE 11.4

The enzyme urease catalyses the hydrolysis of urea, $CO(NH_2)_2$, to carbon dioxide and ammonia:

$$CO(NH_2)_2 + H_2O \rightarrow CO_2 + 2NH_3$$

$$\text{urea + water} \rightarrow \text{carbon dioxide + ammonia}$$

Calculate the enthalpy of this reaction, given that the enthalpies of formation of urea, water, carbon dioxide and ammonia are –333.5, –285.8, –393.5 and –46.1 kJ mol^{-1}, respectively.

ANSWER

First we calculate the total enthalpy of formation of the products:

$$\begin{aligned}\Sigma\, \Delta H_f(\text{products}) &= \Delta H_f(CO_2) + 2\Delta H_f(NH_3)\\ &= -393.5 + 2 \times (-46.1)\\ &= -485.7 \text{ kJ mol}^{-1}\end{aligned}$$

We next calculate the total enthalpy of formation of the reactants:

$$\begin{aligned}\Sigma\, \Delta H_f(\text{reactants}) &= \Delta H_f(CO(NH_2)_2) + \Delta H_f(H_2O)\\ &= -333.5 + (-285.8)\\ &= -619.3 \text{ kJ mol}^{-1}\end{aligned}$$

Finally, we subtract the result for the reactants from the result for the products:

$$\begin{aligned}\Delta H_{reaction} &= \Sigma\, \Delta H_f(\text{products}) - \Delta H_f(\text{reactants})\\ &= -485.7 - (-619.3)\\ &= 133.6 \text{ kJ mol}^{-1}\end{aligned}$$

QUESTION 11.4

Bacteria of the genus *Acetobacter* extract energy from ethanol, C_2H_5OH, by oxidising it to ethanoic acid, CH_3COOH:

$$C_2H_5OH + O_2 \rightarrow CH_3COOH + H_2O$$

ethanol + oxygen → ethanoic acid + water

Calculate the energy available by this process from one mole of ethanol, given that the enthalpies of formation of ethanol, ethanoic acid and water are −277.7, −484.5 and −285.8 kJ mol^{-1}, respectively.

11.9 THE SECOND LAW OF THERMODYNAMICS

The first law of thermodynamics allows us to calculate the energy changes involved in chemical reactions but gives us no information about whether the reaction is likely to proceed or not. It is true that many reactions which occur are exothermic, so we might suppose that reactants undergo reaction in order to achieve a lower energy state. However, this cannot be the whole story, because some reactions happen even though they are endothermic. An example can be seen in *worked example 11.4*. Other examples include many dissolution reactions, such as when potassium chloride dissolves in water:

$$KCl_{(s)} \rightarrow KCl_{(aq)} \qquad \Delta H = +17.2 \text{ kJ mol}^{-1}$$

Similarly, dinitrogen tetroxide, N_2O_4, tends to break down to form nitrogen dioxide, NO_2, as the temperature is raised:

$$N_2O_4 \rightarrow 2NO_2 \qquad \Delta H = +57.2 \text{ kJ mol}^{-1}$$

In order to explain these types of reactions, the second law of thermodynamics introduces a new concept, called **entropy**.

In those endothermic reactions which are observed, it is found that invariably the products are in a more dispersed state than the reactants. In the examples given above, the potassium chloride is dispersed into the liquid phase as it dissolves, and one dinitrogen tetroxide molecule breaks up into two molecules which are then free to separate. Entropy measures this tendency towards dispersal: the greater the dispersal brought about by the change the greater is the entropy change associated with a process.

The second law of thermodynamics states:

> In any spontaneous change, the total entropy of the system and its surroundings must increase.

A spontaneous change is one which takes place without any work being done on the system. The entropy change is given the symbol ΔS, and the second law can be restated as:

> In a spontaneous change, ΔS_{total} must be positive.

The units of entropy are joules per kelvin per mole, J K^{-1} mol^{-1}.

The **system** is the particular part of the world that we are studying. It will often be a test-tube or flask containing reactants but it may be the whole laboratory or the entire earth. We are free to consider anything to be 'the system' provided we can define a boundary around it. The **surroundings** are all the rest of the world lying outside the boundaries of the system.

It is, in fact, the entropy change which determines whether a reaction will proceed or not. The energy changes in a reaction only influence the likelihood of the reaction occurring in so far as they increase or decrease the entropy of the system and its surroundings. For example, an exothermic reaction causes the system and its surroundings to warm up. This increases the average speeds of the molecules in the system and its surroundings, leading to a greater tendency towards dispersal. Exothermic reactions therefore cause an increase in entropy and are therefore likely to proceed.

In an endothermic reaction, the enthalpy change is unfavourable to the reaction, since cooling the system leads to lower mean speeds for the molecules and so a lower tendency towards dispersal. However, if the reaction itself involves dispersal, as in the examples given, then the entropy change brought about by this can be sufficient to overcome the unfavourable energy change and allow the reaction to proceed.

■ 11.10 FREE ENERGY

The effects of enthalpy and entropy changes brought about by a reaction can be summarised in an equation:

$$\Delta G = \Delta H - T\Delta S$$

In this equation, ΔH is the enthalpy change for the system, T is the absolute temperature and ΔS is the entropy change for the system. ΔG is called the **free energy** of the reaction. If ΔG is negative, then the total entropy in the system and surroundings is increased, and the reaction may proceed. If ΔG is positive, then the total entropy is decreased, and the reaction will not proceed.

■ 11.11 INTERACTION OF ΔH WITH $T\Delta S$

Four possibilities exist:

1. ΔH is negative and ΔS is positive. In this case, ΔG will be negative at all temperatures. Such a reaction will be spontaneous at all temperatures.
2. ΔH is negative and ΔS is negative. In this case ΔG will be negative provided $T\Delta S$ is smaller than ΔH. However, $T\Delta S$ increases as temperature increases, so there will be some temperature above which ΔG is positive. The reaction will be spontaneous below this temperature, but not above it.
3. ΔH is positive and ΔS is positive. In this case, ΔG will be negative only if $T\Delta S$ is larger than ΔH. When T is small, therefore, the reaction will not proceed, but as T becomes larger, the reaction will become spontaneous.
4. ΔH is positive and ΔS is positive. In this case, ΔG will be positive at all temperatures and so the reaction will never be spontaneous.

These results are summarised in Table 11.1.

Table 11.1 Relationship between Δ*H*, Δ*S* and Δ*G*

ΔH	ΔS	ΔG
Negative	Positive	Negative at all temperatures
Negative	Negative	Negative at low temperatures
Positive	Positive	Negative at high temperatures
Positive	Negative	Positive at all temperatures

WORKED EXAMPLE 11.5

During aerobic respiration in animals and plants, glucose is oxidised to carbon dioxide and water:

$$C_6H_{12}O_6 + 6O_2 \rightarrow 6CO_2 + 6H_2O$$

$$\text{glucose} + \text{oxygen} \rightarrow \text{carbon dioxide} + \text{water}$$

Calculate the entropy change per mole of glucose at 37°C, given that the enthalpy of the reaction, ΔH, is equal to –2807.8 kJ mol^{-1} and the free energy change, ΔG, is equal to –3089.0 kJ mol^{-1}.

ANSWER

First, we must convert the temperature in degrees Celsius to kelvin:

$$\begin{aligned}\text{Temperature/K} &= \text{Temperature/°C} + 273\\ &= 37 + 273\\ &= 310\end{aligned}$$

The values can then be inserted in the equation:

$$\Delta G = \Delta H - T\Delta S$$

$$-3089.0 = -2807.8 - (310 \times \Delta S)$$

Add 2807.8 to each side of the equation:

$$-3089.0 + 2807.8 = -2807.8 - (310 \times \Delta S) + 2807.8$$

$$-281.2 = -310 \times \Delta S$$

Divide both sides of the equation by –310:

$$\frac{-281.2}{-310} = \frac{-310\Delta S}{-310}$$

$$0.907 = \Delta S$$

$$\Delta S = 0.907 \text{ kJ K}^{-1} \text{ mol}^{-1}$$
$$= 907 \text{ J K}^{-1} \text{ mol}^{-1}$$

Thus the entropy change for the oxidation of glucose at 37°C is +907 J K^{-1} mol^{-1}.

QUESTION 11.5

One step in the TCA cycle involves the addition of water to fumarate, $C_4H_2O_4^{2-}$, to produce malate, $C_4H_4O_5^{2-}$:

$$C_4H_2O_4^{2-} + H_2O \rightarrow C_4H_4O_5^{2-}$$

fumarate + water → malate

Calculate the entropy change involved in this reaction at 25°C, given that $\Delta H = 14.9$ kJ mol^{-1} and $\Delta G = -3.7$ kJ mol^{-1}.

■ SUMMARY

The first law of thermodynamics states that energy can be neither created nor destroyed. This enables us to calculate the enthalpies involved in chemical and biochemical reactions, using Hess's law. The second law of thermodynamics states that for any reaction to occur, the total entropy of system and surroundings must increase as a result of that reaction. This allows us to define a parameter, the free energy, ΔG, which enables us to decide whether a given reaction is possible or not under given circumstances.

■ SUGGESTED FURTHER READING

Atkins, P.W. (1994) *Physical Chemistry*, Chs 4 and 5, 5th edn. Oxford University Press.

■ END OF CHAPTER QUESTIONS

Question 11.6 (a) State the first law of thermodynamics.
(b) What is the SI unit of energy?

Question 11.7 Explain what is meant by:
(a) exothermic,
(b) endothermic.
If, during a reaction, the reaction vessel gets hot, which type of reaction is occurring?

Question 11.8 In part of the TCA cycle, glucose is converted to pyruvic acid:

$$C_6H_{12}O_6 + O_2 \rightarrow 2C_3H_4O_3 + 2H_2O$$

glucose + oxygen → pyruvic acid + water

Calculate the enthalpy change associated with this reaction, given the following enthalpies of combustion:

$$C_6H_{12}O_6 + 6O_2 \rightarrow 6CO_2 + 6H_2O \qquad \Delta H_{comb} = -2822 \text{ kJ mol}^{-1}$$

$$C_3H_4O_3 + \tfrac{5}{2}O_2 \rightarrow 3CO_2 + 2H_2O \qquad \Delta H_{comb} = -1168 \text{ kJ mol}^{-1}$$

Question 11.9 Fumaric and maleic acids are a pair of geometric isomers:

```
H       COOH          HOOC     COOH
  \    /                  \    /
   C=C                     C=C
  /    \                  /    \
HOOC    H                H      H
 fumaric acid           maleic acid
```

The enthalpies of formation of fumaric and maleic acids are −810 and −785 kJ mol^{-1} respectively. Calculate the enthalpy of the isomerisation reaction that converts fumaric to maleic acid.

REACTIONS AND EQUILIBRIUM

12

12.1 INTRODUCTION

In this chapter we continue the investigation of chemical reactivity as it relates to the biochemical reactions necessary for the growth, reproduction, repair and maintenance of living cells. Many of these reactions are equilibrium reactions, that is, reactions that do not go to completion. The role of free energy, ΔG, in determining equilibrium constants is examined. Also important to living organisms is the rate at which a reaction proceeds. This is controlled by the activation energy of the reaction. Organisms use enzymes to modify the activation energies of reactions and so control the rates of these reactions.

12.2 ΔG AND EQUILIBRIUM

In Chapter 11 we saw that the value of ΔG allowed us to decide whether a reaction was feasible or not. To summarise what was said there: if the value of ΔG is negative, then the reaction can proceed; if the value is positive, it cannot. However, we saw in Chapter 4 that not all reactions go to completion; many reactions reach an equilibrium position in which both reactants and products are present. What is the value of ΔG for these reactions?

It turns out that there is a simple relationship between ΔG and the equilibrium constant, K_{eq}:

$$\Delta G = -RT \ln K_{eq}$$

where R is a constant called 'the molar gas constant' or 'the universal gas constant' and has the value 8.314 J K^{-1} mol^{-1} and T is the temperature in Kelvin. The derivation of this equation is beyond the scope of this text but can be found in the books suggested in section 12.11.

Using this equation, we can investigate the effect of different ΔG values on the equilibrium constant.

WORKED EXAMPLE 12.1

The free energy change associated with the complete oxidation of glucose:

$$C_6H_{12}O_{6(s)} + 6O_{2(g)} \rightarrow 6CO_{2(g)} + 6H_2O_{(l)}$$

glucose + oxygen → carbon dioxide + water

is –3089.0 kJ mol^{-1}. What is the equilibrium constant for this reaction at 37°C?

ANSWER

First, convert the temperature to kelvin:

$$37°C = 37 + 273\ K$$
$$= 310\ K$$

Using the equation

$$\Delta G = -RT \ln K_{eq}$$

$$-3089.0 \times 10^3 = -8.314 \times 310 \times \ln K_{eq}$$

$$\ln K_{eq} = \frac{-3089.0 \times 10^3}{-8.314 \times 310}$$

$$= 1198.5$$

$$K_{eq} = 3.18 \times 10^{520}$$

Reactions with large negative values of ΔG go to completion

This is an enormous number. Bearing in mind that the equilibrium constant for this reaction would be written:

$$K_{eq} = \frac{[CO_2]^6[H_2O]^6}{[C_6H_{12}O_6][O_2]^6}$$

the value of the equilibrium constant shows that when equilibrium is reached, essentially all the glucose and oxygen will have been converted into carbon dioxide and water. The reaction is said to have gone to completion.

WORKED EXAMPLE 12.2

The free energy change for the addition of water to fumarate, $C_4H_2O_4^{2-}$, to produce malate, $C_4H_4O_5^{2-}$

$$C_4H_2O_4^{2-} + H_2O \rightarrow C_4H_4O_5^{2-}$$

fumarate + water → malate

is –3.7 kJ mol^{-1}. What is the equilibrium constant for this reaction at 37°C?

ANSWER

$$37°C = 310\ K$$

Using the equation

$$\Delta G = -RT \ln K_{eq}$$

$$-3.7 \times 10^3 = -8.314 \times 310 \times \ln K_{eq}$$

$$\ln K_{eq} = \frac{-3.7 \times 10^3}{-8.314 \times 310}$$

$$= 1.44$$

$$K_{eq} = 4.20$$

The equilibrium constant for this reaction would be written:

Reactions with small values of ΔG reach equilibrium

$$K_{eq} = \frac{[C_4H_4O_5^{2-}]}{[C_4H_2O_4^{2-}][H_2O]}$$

and its value, 4.20, indicates that, at equilibrium, a substantial amount of reactants will still be present. This is typical of reactions for which ΔG is small, either negative or positive.

As a rough guide, it can be stated that:

(a) Reactions with ΔG values more negative than about -10 kJ mol^{-1} go to completion. At this value about 98 per cent of the reactants are converted to products when equilibrium is reached.
(b) Reactions with ΔG values more positive than $+10$ kJ mol^{-1} do not go at all. At this value only about 2 per cent of reactants are converted to products when equilibrium is reached.
(c) Reactions with ΔG values between these two values come to equilibrium with significant amounts of both reactants and products present.

QUESTION 12.1

The two amino acids leucine, $(CH_3)_2CHCH(NH_3^+)COO^-$, and glycine, $CH_2(NH_3^+)COO^-$, can combine to form the dipeptide leucylglycine:

$$(CH_3)_2CHCH(NH_3^+)COO^- + CH_2(NH_3^+)COO^- \rightleftharpoons (CH_3)_2CHCH(NH_3^+)CONHCH_2COO^- + H_2O$$

$$\text{leucine} + \text{glycine} \rightleftharpoons \text{leucylglycine} + \text{water}$$

The value of ΔG for this reaction is $+13.0$ kJ mol^{-1} at 37°C.

(a) Calculate the equilibrium constant for this reaction.
(b) Comment on the relative concentrations of reactants and products in the equilibrium mixture.

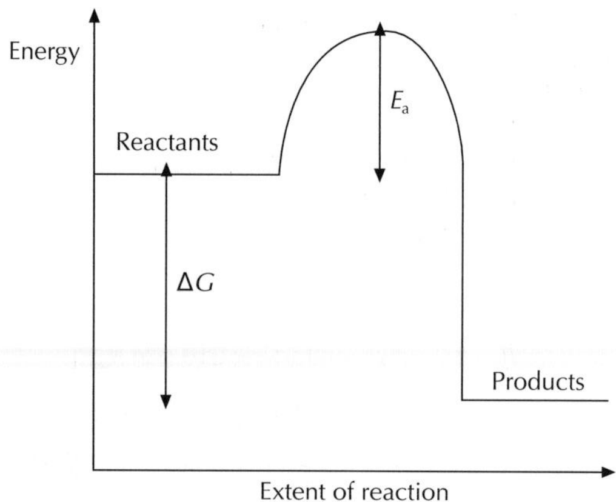

• **Figure 12.1** Typical reaction profile.

■ 12.3 ACTIVATION ENERGY

In the previous section we saw that how far a reaction goes, whether to completion or to some equilibrium position, depends on the size and sign of ΔG. It is important to realise that this gives us no information at all about how quickly the reaction will go. For example, in the previous section we found that ΔG for the reaction between glucose and oxygen was extremely large and negative, indicating that the reaction would go to completion. However, we know that glucose can be safely stored in contact with air (and hence with oxygen) for very long periods of time with no apparent reaction. This occurs because, in order to start the reaction, a lot of energy has to be supplied, and this energy is not normally available in the environment. The energy which must be supplied to the reactants in order for them to react is called the **activation energy**, E_a. This energy is required to stretch or otherwise deform bonds in the reactant molecules so that these bonds become vulnerable to attack. The relationship between free energy of reaction and activation energy can be depicted in a diagram (see Figure 12.1).

In Figure 12.1, we can see that the products are at a lower energy level than the reactants, indicating that ΔG is negative for this reaction. However, before the reactants can be transformed into products, an amount of energy E_a must be supplied to them before reaction can proceed – there is an energy barrier to be overcome. This can be done either by supplying energy to the reactants, usually by heating them, or by finding some means of lowering the energy barrier.

■ 12.4 THE EFFECT OF TEMPERATURE ON REACTION RATE

Increasing the temperature always increases reaction rates

Increasing the temperature of a substance causes the particles comprising the substance to move more quickly, hence increasing their kinetic energy. The faster the molecules are moving, the more likely it is that, when they collide, they have enough energy to surmount the energy barrier and thus react to form products. With molecules travelling at greater speeds, more collisions occur every second and this also increases the rate of reaction. The effect that raising the temperature has on the rate of a reaction is quite marked. A rough rule of thumb is that increasing the temperature by 10°C doubles the rate of reaction. This comes about because of the effect of temperature on the distribution of molecular speeds. This is illustrated in Figure 12.2, which shows two effects of increasing temperature:

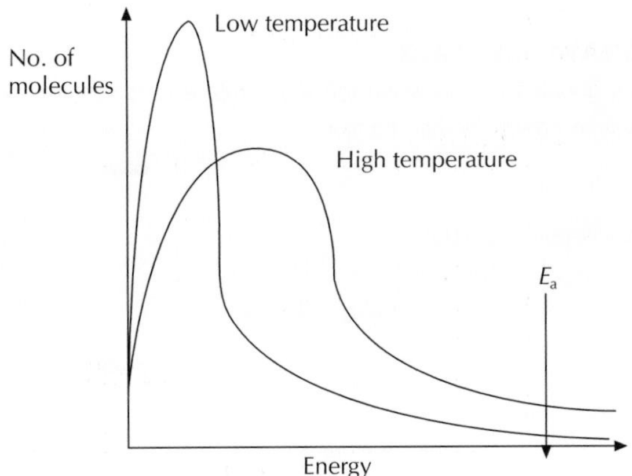

• **Figure 12.2** Effect of increasing temperature on the energy of molecules.

1. As the temperature is raised, the average energy of the molecules increases. The shift in the peak of the distribution to the right and thus to higher energy illustrates this.
2. More importantly, as the temperature is raised, the number of molecules with much higher than average energies increases much faster than the average energy. It is these molecules, with energies above E_a, that cause the reaction rate to increase rapidly with increasing temperature.

12.5 ARRHENIUS'S EQUATION

The effect of temperature on reaction rates is modelled by Arrhenius's equation:

$$\text{Rate} = Ae^{-\frac{E_a}{RT}}$$

where A is called the Arrhenius factor and is related to the number of collisions a molecule makes in each second, e is the base of natural logarithms, approximately equal to 2.718, E_a is the activation energy, R the molar gas constant, 8.314 J K^{-1} mol^{-1}, and T the temperature in Kelvin. Taking natural logarithms of both sides of this equation gives:

$$\ln(\text{Rate}) = \ln A - \frac{E_a}{RT}$$

If we measure the rate of reaction at two different temperatures, T_1 and T_2, this equation allows us to calculate the activation energy of the reaction:

$$\ln\left(\frac{\text{Rate}_1}{\text{Rate}_2}\right) = -\frac{E_a}{R}\left(\frac{1}{T_1} - \frac{1}{T_2}\right)$$

where Rate_1 and Rate_2 are the rates of reaction at temperatures T_1 and T_2, respectively.

WORKED EXAMPLE 12.3

A certain reaction doubles in rate when the temperature is raised from 298 K to 308 K. What is the activation energy of this reaction?

ANSWER

If the reaction rate is doubled, then

$$\text{Rate}_2 = 2\text{Rate}_1$$

and so

$$\ln\left(\frac{\text{Rate}_1}{\text{Rate}_2}\right) = \ln\frac{1}{2}$$

Using the equation

$$\ln\left(\frac{\text{Rate}_1}{\text{Rate}_2}\right) = -\frac{E_a}{R}\left(\frac{1}{T_1} - \frac{1}{T_2}\right)$$

and substituting in the given values gives

$$\ln\left(\frac{1}{2}\right) = -\frac{E_a}{R}\left(\frac{1}{298} - \frac{1}{308}\right)$$

Since

$$\ln\left(\frac{1}{2}\right) = -0.693,$$

$$\frac{1}{298} - \frac{1}{308} = 1.09 \times 10^{-4}\ \text{K}^{-1}$$

and

$$R = 8.314\ \text{J K}^{-1}\ \text{mol}^{-1}$$

$$-0.693 = -\frac{E_a}{8.314}(1.09 \times 10^{-4})$$

Rearranging this gives:

$$-E_a = \frac{-0.693 \times 8.314}{1.09 \times 10^{-4}}$$

$$E_a = 52\,900\ \text{J mol}^{-1}$$
$$= 52.9\ \text{kJ mol}^{-1}$$

QUESTION 12.2

The rate of a reaction increases from $1.24 \times 10^{-3}\ \text{s}^{-1}$ at 295 K to $3.16 \times 10^{-3}\ \text{s}^{-1}$ at 303 K. Calculate the activation energy of this reaction.

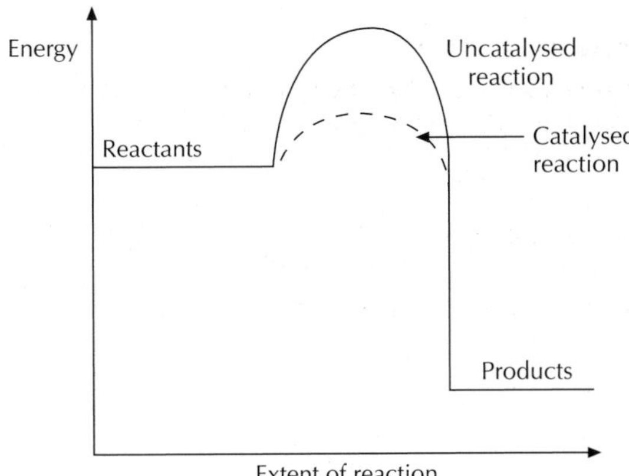

• **Figure 12.3** Effect of a catalyst on activation energy.

■ 12.6 CATALYSIS

Increasing temperature increases reaction rate by increasing the number of molecules with sufficient energy to overcome the activation barrier. A similar effect can be produced by reducing the activation barrier. This is what catalysts do. The effect is illustrated in Figure 12.3.

Catalysts increase the rate of a reaction by lowering the activation energy

The catalyst lowers the activation energy of a reaction, so that more molecules will have the necessary energy to react. A catalyst does nothing to change the position of an equilibrium, which is determined by the free energy difference between reactants and products; it merely changes the rate at which equilibrium is attained.

Many catalysts produce this effect by offering an alternative route from reactants to products. This involves one or more of the reactants binding to the catalyst surface. Interaction between the reactant(s) and the catalyst surface leads to a weakening of the bonds in the reactant, making them more likely to react.

■ 12.7 ENZYME CATALYSIS

Nearly all biochemical reactions occurring in living organisms are controlled by enzymes, which are biological catalysts. Enzymes are protein molecules and their activity depends on the folding of the peptide chain or chains which comprise the protein. The enzyme is folded in such a way as to provide a binding site, called the **active site**, for the target molecule, or **substrate**, of the enzyme. This binding is highly specific and a given enzyme will usually react with only a single substrate or a small class of closely related substrates. Thus, D-glucose oxidase will only bind to D-glucose and not to L-glucose or any other sugar, while alcohol dehydrogenase will bind to a number of low-molecular-weight alcohols but not to any other type of molecule.

A simple model for the mechanism of an enzyme reaction is illustrated in Figure 12.4.

The substrate interacts with the active site of the enzyme, which is exactly the right shape and size to fit the substrate, and provides sites to which the substrate can bond. The resulting combination of enzyme and substrate is called the enzyme–substrate complex. Reaction, in this case a bond fission, then takes place. Finally, the enzyme releases the products and is ready to accept another substrate molecule.

In reality, the interactions are rather more complicated than illustrated in this example. Both the enzyme and the substrate change shape somewhat as the initial binding takes place. It is this enforced alteration of the shape of the substrate that causes bonds within it to become strained and likely to react, while the change in shape of the enzyme

• **Figure 12.4** Mechanism of an enzyme-catalysed reaction.

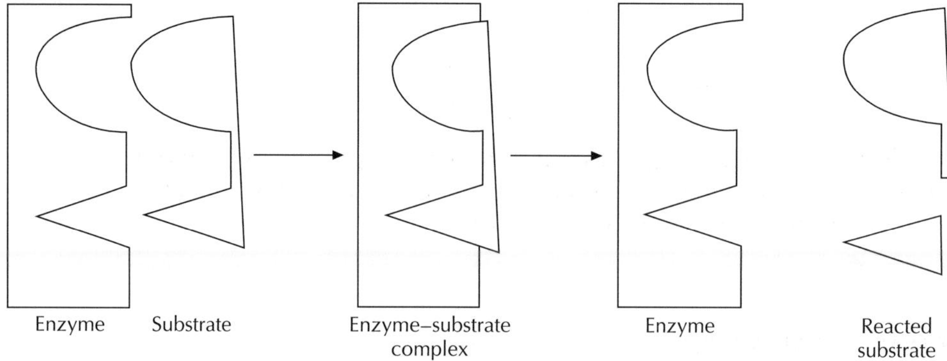

serves to bring close to the substrate side chain or prosthetic groups which will cause reaction to take place. Once the reaction has occurred, the product molecules no longer fit the active site and are released.

■ 12.8 KINETICS OF ENZYME REACTIONS

The process of enzyme catalysis can be modelled by the following sequence of reactions:

$$E + S \rightleftharpoons ES \rightarrow E + P$$

Here, E represents the enzyme, S the substrate, ES the enzyme–substrate complex and P the products of the reaction. This sequence can be analysed mathematically (see Appendix, Derivation 12.1) to obtain an equation for the rate of an enzyme reaction in terms of substrate concentration and two constants which are characteristic of the enzyme being used:

$$\text{Rate} = \frac{V_{max}[S]}{K_m + [S]}$$

This is called the Michaelis–Menten equation. [S] represents the concentration of the substrate. V_{max} and K_m are constants. V_{max} is the maximum rate at which the enzyme can work. This is the rate of the reaction when the substrate is present in large excess, so that each enzyme molecule is supplied with another molecule of substrate as soon as the previous one has reacted and dissociated from the enzyme. Under these conditions the limiting factor is usually the rate at which diffusion can supply substrate to the enzyme. K_m is called the **Michaelis constant** and it provides a measure of the strength of the bonding of the substrate and the products to the enzyme.

■ 12.9 FINDING K_m AND V_{max}

The Michaelis–Menten equation can be rewritten as follows:

$$\text{Rate} = \frac{V_{max}[S]}{K_m + [S]}$$

$$\frac{1}{\text{Rate}} = \frac{K_m + [S]}{V_{max}[S]}$$

The fraction on the right-hand side can be split into two:

$$\frac{1}{\text{Rate}} = \frac{K_m}{V_{max}} \cdot \frac{1}{[S]} + \frac{1}{V_{max}}$$

A graph of 1/Rate versus 1/[S] will give a straight line of intercept $1/V_{max}$ and gradient K_m/V_{max}. Thus, if we measure the rate of the reaction for a range of substrate concentrations we can plot this graph and obtain values of K_m and V_{max}.

WORKED EXAMPLE 12.4

Carbon dioxide is hydrated by the enzyme carbonic anhydrase to form the bicarbonate, HCO_3^-, ion:

$$CO_2 + H_2O \rightarrow HCO_3^- + H^+$$

carbon dioxide + water → bicarbonate ion + hydrogen ion

The following rate data were gathered for this enzyme:

$[CO_2]$/mmol dm^{-3}	0.76	1.51	3.78	7.57	15.1
Initial rate/mmol dm^{-3} s^{-1}	0.0062	0.0116	0.0204	0.0287	0.0368

Find K_m and V_{max} for this reaction.

ANSWER

First, the data must be transformed into the form required for the graph, i.e. $1/[CO_2]$ and 1/Rate.

$1/[CO_2]$/mmol^{-1} dm^3	1.316	0.6623	0.2546	0.1321	0.06623
1/Rate/mmol^{-1} dm^3 s	161.3	86.21	49.02	34.84	27.17

Then a graph must be plotted of 1/Rate versus $1/[CO_2]$:

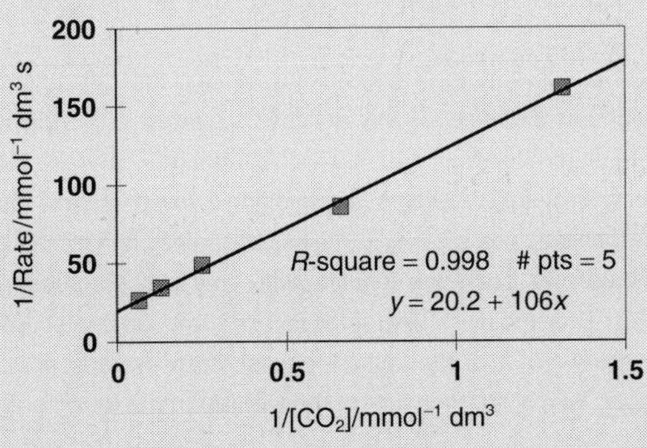

The graph tells us that the intercept = 20.2 and the gradient = 106.

$$\frac{1}{V_{max}} = 20.2$$

$$V_{max} = \frac{1}{20.2}$$

$$= 0.050 \text{ mmol dm}^{-3} \text{ s}^{-1}$$

$$\frac{K_m}{V_{max}} = 106$$

$$K_m = 106 \times V_{max}$$

$$= 106 \times 0.050$$

$$= 5.25 \text{ mmol dm}^{-3}$$

QUESTION 12.3

Penicillinase is an enzyme which hydrolyses penicillin. Penicillin solutions of various concentrations were made up and reacted with the same quantity of penicillinase. The following results were obtained:

$[\text{Penicillin}]_0/10^{-5}$ mol dm^{-3}	Initial rate/10^{-8} mol dm^{-3} min^{-1}
0.1	1.10
0.3	2.50
0.5	3.40
1.0	4.50
3.0	5.80
5.0	6.10

Find K_m and V_{max} for penicillinase under these conditions.

■ SUMMARY

All reactions are equilibrium reactions, in which the position of equilibrium is determined by the value of ΔG. However, when ΔG is large and negative the equilibrium lies so far towards the products that the reaction can be considered to go to completion, and when ΔG is large and positive the equilibrium lies so much towards the reactants that it can be said that the reaction does not go at all. Only when ΔG is small (within about 10 kJ mol^{-1} of zero) will there be significant amounts of both reactants and products when the reaction comes to equilibrium.

While the position of equilibrium is determined by ΔG, the rate at which equilibrium is attained is controlled by the size of the activation energy, E_a. This energy has to be expended in order to distort bonds in such a way that reaction can take place. Catalysts can lower the activation energy for a reaction by providing a lower-energy route from reactants to products. Living organisms use catalytic proteins called enzymes for this purpose. Enzymes are highly specific catalysts, each one reacting with only a single substrate or small group of similar substrates. Enzymes are characterised by two constants determined from the Michaelis–Menten equation: V_{max}, the fastest rate at which the enzyme can work, and K_m, the Michaelis constant, which is a measure of the stability of the enzyme–substrate complex.

■ SUGGESTED FURTHER READING

Atkins, P.W. (1994) *Physical Chemistry*, Chs 5 and 25, Oxford University Press.
Lehninger, A.L., Nelson, D.L. and Cox, M.M. (1997) *Principles of Biochemistry*, Ch. 8, Worth, New York.

■ END OF CHAPTER QUESTIONS

Question 12.4 ATP can be hydrolysed to produce ADP and inorganic phosphate, P_i. ΔG for this reaction is –30.9 kJ mol^{-1}. Calculate the equilibrium constant, K_{eq}, for this reaction at 37°C.

Question 12.5 The enzyme glucophosphomutase brings about the rearrangement of glucose-1-phosphate to glucose-6-phosphate. The equilibrium constant for this reaction = 19. What is ΔG for this reaction at 37°C?

Question 12.6 Explain why raising the temperature increases the rate of a reaction.

Question 12.7 The hydrolysis of sucrose proceeds as follows:

$$C_{12}H_6O_{12} + H_2O \rightarrow C_6H_{12}O_6 + C_6H_{12}O_6$$

$$\text{sucrose} + \text{water} \rightarrow \text{glucose} + \text{fructose}$$

The rate constant for this reaction is 6.1×10^{-5} s^{-1} at 298 K in acidified water. At 310 K the rate constant = 3.2×10^{-4} s^{-1}. What is the activation energy of this reaction?

Question 12.8 Describe the mechanism by which enzymes catalyse reactions, using suitable diagrams.

Question 12.9 The enzyme isocitrate lyase catalyses the following reaction:

$$\text{isocitrate} \rightarrow \text{glyoxylate} + \text{succinate}$$

The following results were obtained when the enzyme reaction was studied at a temperature of 32°C:

Concentration of isocitrate /10^{-5} mol dm^{-3}	Initial rate of reaction /10^{-9} mol dm^{-3} min^{-1}
1.0	2.86
2.0	4.21
3.0	5.00
4.0	5.52
5.0	5.88

Calculate (a) V_{max} and (b) K_m for this enzyme.

13 LIGHT

13.1 INTRODUCTION

An understanding of light is important in order to appreciate its effect on many living systems. Light reflected from objects into our eyes enables us to see the object because molecules in the eye can absorb light. Light absorbance allows us to appreciate the beauty of great works of art as well as providing us with a means by which to perceive the environment around us. Plants are able to use light-absorbing pigments to trigger seasonal changes in development such as flowering. In an energetic sense light is important. The absorbance of light by plants during photosynthesis provides the energy that drives the synthesis of complex biomolecules.

13.2 LIGHT IS PART OF THE ELECTROMAGNETIC SPECTRUM

Light is part of the electromagnetic spectrum

Light is part of a much wider range of emissions of energy that is called the electromagnetic spectrum. The electromagnetic spectrum covers the emission of light from X-rays through to radio waves. A diagram showing the position of visible light in the electromagnetic spectrum is given in Figure 13.1. Figure 13.1 shows that different types of electromagnetic radiation have different wavelengths. For example, X-rays have very short

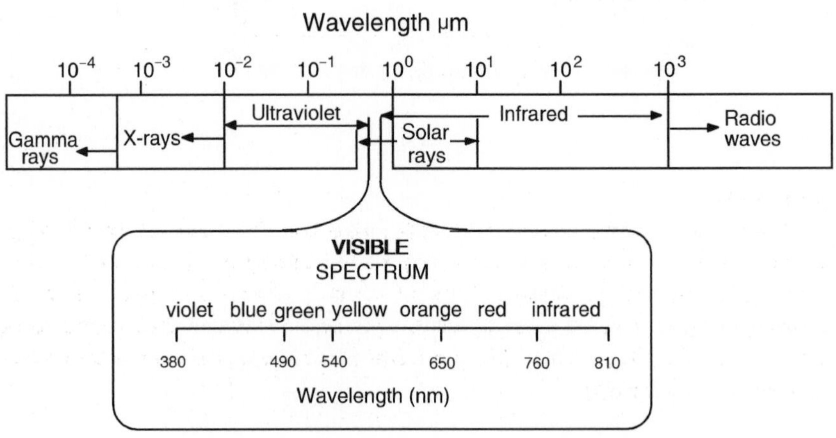

• **Figure 13.1** The electromagnetic spectrum expanded in the region containing visible light.

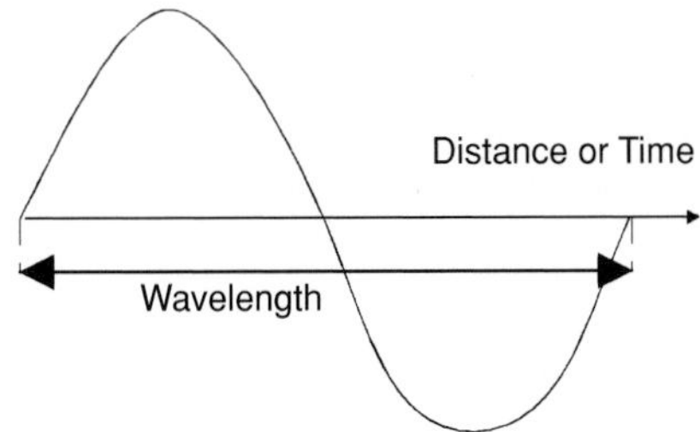

• **Figure 13.2** A sine wave showing the wavelength marked between similar points in consecutive waves.

wavelengths whilst radio waves can have wavelengths of over 1500 m. A better understanding of light (and therefore electromagnetic radiation) can be achieved if we examine what is meant by wavelength and frequency in more detail.

■ 13.3 WAVELENGTH AND FREQUENCY

Light will be better understood if we examine some features of a simple sine wave (Figure 13.2). A wave can be thought of as any shape that repeats at regular intervals along a path. The sine wave shown in Figure 13.2 repeats every time that the wave rises through the *x*-axis. The length between rises is called the wavelength. The symbol for wavelength is the Greek letter lambda (λ). A sine wave will keep repeating indefinitely and so the wavelength is an important way of defining that waveform. In the sine wave shown in Figure 13.2 the wavelength could be measured between any two successive repeating parts.

WORKED EXAMPLE 13.1

Figure 13.3 shows a sine wave. On the sine wave there are several distances marked with letters. Identify which letters correspond to the wavelength.

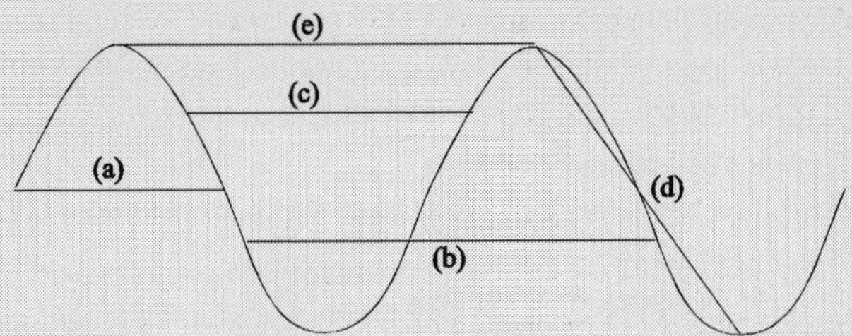

• **Figure 13.3** Sine wave for worked example 13.1.

ANSWER

The letters which correspond to the wavelength are **b** and **e**. This is because the wavelength can be measured between any two repeating points on a sine wave. Thus **b** and **e** are both wavelengths measured at different points on the sine wave; **a** and **c** are not, as the distance is only about half of a wavelength; **d** is not, as the waveform does not repeat until the next pattern is fully repeated. The two points at the end of line **d** represent opposite extremes of the waveform.

QUESTION 13.1

Figure 13.4 shows a waveform. There are several distances identified with letters. Identify which letters correspond to distances that are the wavelength.

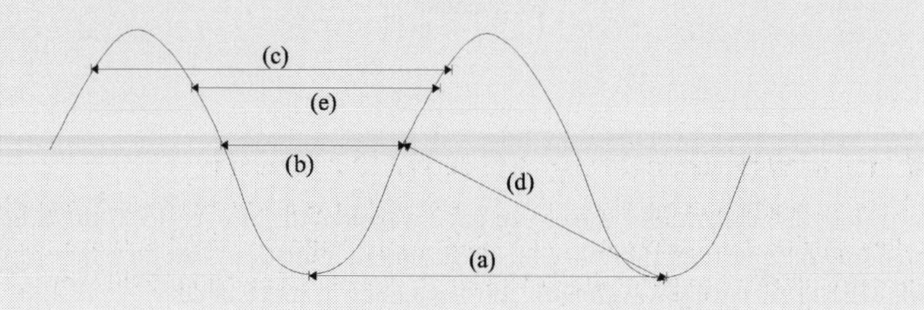

• **Figure 13.4** Sine wave for question 13.1.

Light travels distances at a set speed. It is well known that light takes about eight minutes to travel from the sun to earth. The *x*-axis could also be drawn using time instead of distance. The sine wave will then repeat a number of times in a second, a value that is called the frequency. The frequency has the symbol of the Greek letter nu which looks like an italicised v (ν). The speed of light (symbol *c*) is a constant for any defined medium and it follows that wavelength and frequency are inversely related. The larger the wavelength the smaller the frequency and vice versa. This relationship can be written as:

$$\nu\lambda = c \tag{13.1}$$

WORKED EXAMPLE 13.2

Calculate the frequency of light of wavelength 550 nm, given $c = 3 \times 10^8$ m s^{-1}.

ANSWER

Convert the wavelength to metres: 550 nm = 5.5×10^{-7} m

Transform Equation 13.1 to isolate the term v. To do this divide both sides of Equation 13.1 by λ and cancel out terms. This gives $\nu = c/\lambda$.

Substitute terms into the transformed equation.

$$\nu = \frac{3 \times 10^8}{5.5 \times 10^{-7}}$$

$$= 5.45 \times 10^{14}\ \text{s}^{-1}$$

QUESTION 13.2

Calculate the frequency of light of a wavelength in the near-UV of 254 nm.

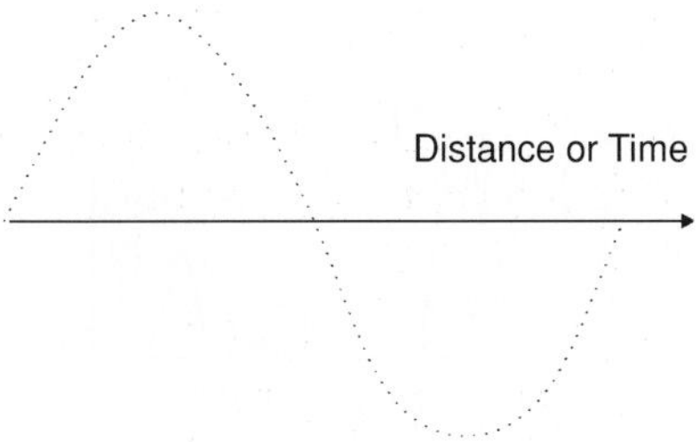

• **Figure 13.5** The dual nature of light showing both wave and quantum nature.

■ 13.4 THE QUANTUM THEORY OF LIGHT

Electromagnetic energy also behaves as though it is a stream of small packets of energy called quanta. The quanta are called photons. In diagrams we often represent a photon by an arrow with a squiggly tail. A convenient way to think about light would be to divide the waveform into lots of tiny particles behaving like a wave as shown in Figure 13.5. The greater the frequency of the light the greater the energy of each photon (E). This can be expressed as:

The symbol for a photon

$$E = h\nu \qquad (13.2)$$

where E is the energy, ν is the frequency and h is a constant, called the Planck constant (value 6.62×10^{-34} J s^{-1}), that relates the two. It is now possible to combine the two equations by substituting the frequency term in Equation 13.2 by E/h to give:

$$E = \frac{hc}{\lambda} \qquad (13.3)$$

In this equation the smaller the wavelength of the light the higher its energy. This equation has a direct application to photosynthesis in that photosynthetic reaction centres can use light of up to a certain wavelength (and therefore energy) to carry out light reactions.

WORKED EXAMPLE 13.3

Calculate the energy of a photon of light of wavelength 550 nm.

ANSWER

Convert the wavelength to metres: 550 nm = 5.5×10^{-7} m

Substitute terms into Equation 13.3:

$$E = \frac{(6.64 \times 10^{-34}) \times (3 \times 10^{8})}{5.5 \times 10^{-7}}$$

$$= 3.62 \times 10^{-19} \text{ J}$$

QUESTION 13.3

(a) Calculate the energy of a photon of light of wavelength 198 nm.

(b) Is the energy of a photon of 198 nm greater than that of a photon of 550 nm?

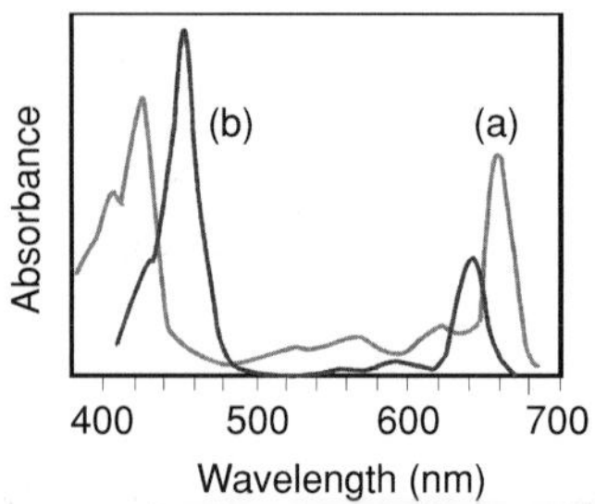

• **Figure 13.6** The absorption spectra of chlorophylls a and b.

Table 13.1 The relationship between orbital, electron arrangement and covalent bond between carbon atoms

Type of orbital	*Bond between carbons*	*Bond resulting*	*Antibonding orbital*
s or end-on p (p_x)	Single	σ	σ*
p_y or p_z	Double or triple	π	π*

Ultraviolet plus visible light is shortened to UV–visible

13.5 THE ABSORPTION OF LIGHT

Light (and therefore energy) is absorbed by a specific part of a molecule called a chromophore. Different molecules absorb light at different wavelengths. This is shown in Figure 13.6 which shows the visible spectrum of two different chlorophylls that are found in the chloroplast membrane. The molecules absorb differently because of differences in their structure. It is the electronic arrangement of a molecule that is responsible for the absorption of light in the ultraviolet (UV) and visible parts of the electromagnetic spectrum. Absorption of light by electrons held in bonds can be viewed in terms of an energy diagram. The diagram in the margin shows the photon as a squiggly arrow. The energy level of the molecule has jumped as a result of the absorption of a photon. One of the bonds in the molecule has absorbed this energy.

Bonding between the carbon atoms in biological molecules is responsible for many of the light absorptions in the UV–visible region. In Chapters 1 and 2 the arrangement of electrons in covalent bonds was reviewed. Some important facts relating to bonds in organic molecules are summarised in Table 13.1. The two types of bonds that need consideration are σ bonds, which result from the sharing of electrons from the s or p_x orbitals (single carbon–carbon bonds), and π bonds, which result from sharing of p orbitals (forming the second and third bonds). Each bond consists of two electrons with complementary spins. Absorption of a photon of light by electrons causes the electrons within a bond to adopt non-complementary spins. Bonding with non-complementary spins is called an antibonding. The symbols for antibonding σ and π orbitals are σ* and π*. Absorption of a photon of light causes an electron in the highest filled molecular orbital to jump into a higher energy, usually antibonding, orbital. The energy jump requires the absorption of exactly the correct amount of energy. The energy is required to separate the two complementary electrons and to allow one of the separated electrons to jump into the lowest unfilled molecular orbital.

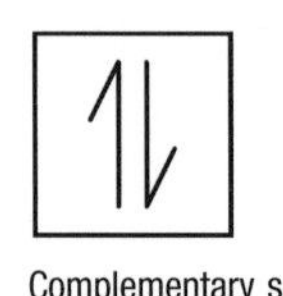

Complementary spins

Equation 13.3 states that lower energy jumps result from higher wavelength light. π to π* transitions require less energy than σ to σ*. Many molecules absorbing visible light have conjugated double-bond systems. Conjugation lowers the energy required for π to π* transitions. The energy jump becomes lower as the degree of conjugation increases.

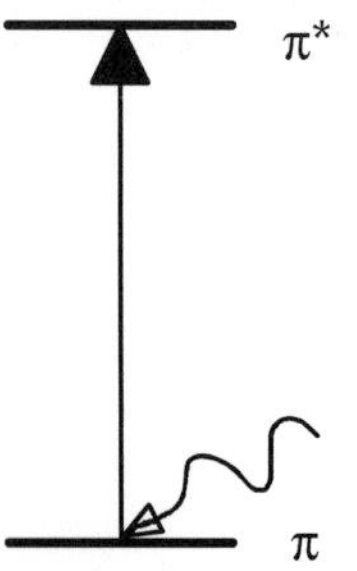

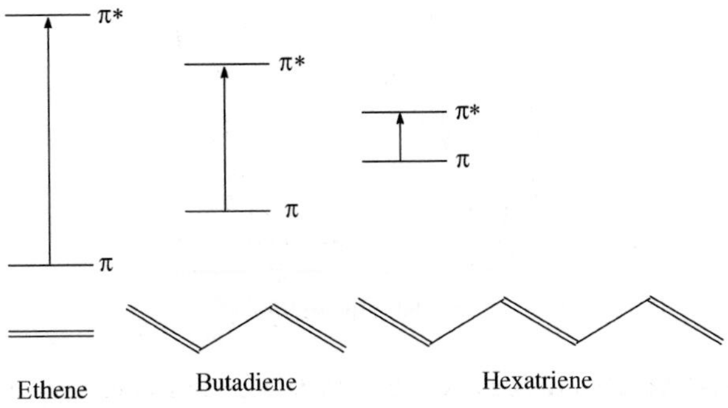

• **Figure 13.7** The effect of degree of conjugation on π to π* transitions.

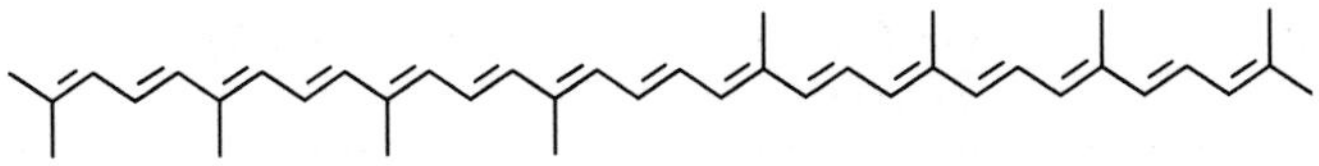

• **Figure 13.8** Lycopene, the red pigment from tomatoes, has an extended conjugated double bond structure.

This is shown in Figure 13.7. Lycopene (Figure 13.8), the red pigment in tomatoes, contains a highly conjugated chain structure, and so absorbs light at the higher end of the visible spectrum.

WORKED EXAMPLE 13.4

Examine Figure 13.9. Explain which compound you would expect to absorb light at the highest wavelength.

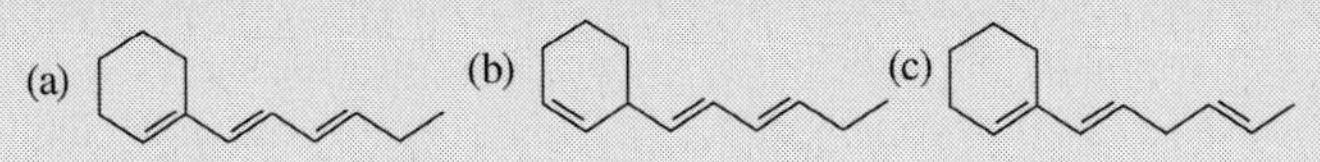

• **Figure 13.9** Three alkenes for worked example 13.4.

ANSWER

Compound (a) has a three-bond conjugated system whilst compounds (b) and (c) have two bonds in conjugation. Therefore, compound (a) will have the lowest energy for a π to π* transition and will absorb light at the highest wavelength.

QUESTION 13.4

Examine Figure 13.10. Which compound would you expect to absorb light at the highest wavelength?

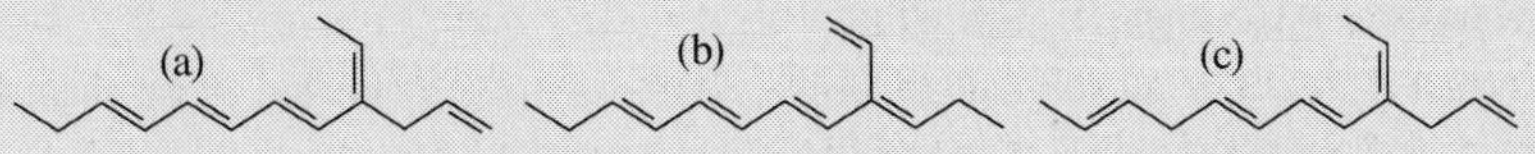

• **Figure 13.10** Three alkenes for question 13.4.

The simplistic explanation for light absorption described above should result in a very narrow absorption band for molecules. The spectra in Figure 13.6 show that molecules can have wide absorption bands. This is because molecules in solution will differ from each other slightly due to a number of complex factors such as bond vibration, polarity, etc. Other groups, particularly those containing unpaired electrons, also contribute to

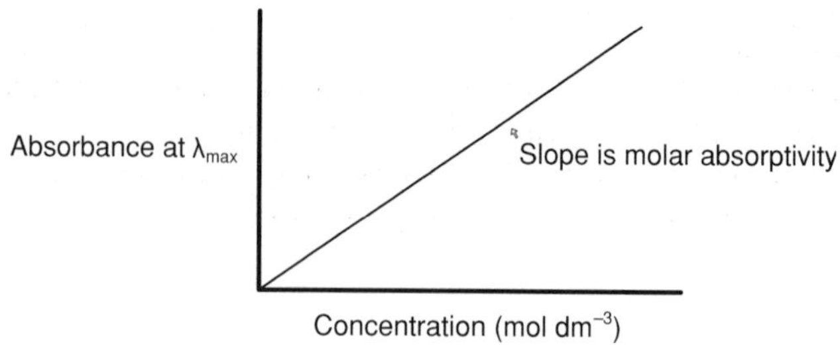

• **Figure 13.11** The relationship between absorbance and wavelength demonstrated by the Beer–Lambert law.

light absorption in the UV–visible region. Thus information about molecular structure and environment can be derived from a consideration of UV–visible spectra.

■ 13.6 THE RELATIONSHIP BETWEEN LIGHT ABSORPTION AND CONCENTRATION

The information that can be derived from study of spectra would be of limited value were it not for the Beer–Lambert law that relates light absorption to concentration (see Appendix, Derivation 13.1):

$$A = \varepsilon c l \tag{13.4}$$

The Beer–Lambert law states that light absorbance (A) at a particular wavelength is proportional to the concentration of the molecule in solution (c). l is the distance that the light has to pass through in the solution being measured; this is normally set at 1 cm so that calculations are simplified. ε is called the molar absorptivity or molar extinction coefficient, a theoretical value for the absorbance of a 1 mol dm^{-3} solution at the wavelength being used for measurements with a 1 cm path length. The wavelength chosen for measurements is normally one where the molecule of interest shows strong absorption characterisics such as a peak wavelength (λ_{max}). This is because the molecule will be easier to measure at low concentrations, as the absorbance will be greater than at other wavelengths. The relationship is shown in graphical form in Figure 13.11. The Beer–Lambert law is useful because many biomolecules absorb in the UV–visible region and their concentration can be measured directly if the molar absorptivity is known. In addition, many molecules, such as proteins, do not absorb in the visible region. Their concentration can be measured after appropriate chemical reactions have been carried out, by using a method known as colorimetry. This is what happens in the colorimetric measurement of proteins in the Biuret reaction.

WORKED EXAMPLE 13.5

Examine Figure 13.6. Explain which visible wavelength would be the most suitable to measure the absorbance of a pure solution of chlorophyll a.

ANSWER

The wavelength at which chlorophyll a shows the highest absorbance is about 420 nm. Measuring the absorbance at this wavelength will show the greatest absorbance change in the visible region. Choosing 660 nm, although a peak wavelength (λ_{max}), would not be suitable as chlorophyll a does not absorb as strongly at this peak as at the peak at 420 nm.

QUESTION 13.5

(a) Examine Figure 13.6. Explain which visible wavelength would be the most suitable to measure the absorbance of a pure solution of chlorophyll b.

(b) How would the choice of wavelengths for measuring chlorophyll b differ if the solution contained unknown amounts of chlorophyll b and chlorophyll a?

WORKED EXAMPLE 13.6

A solution of NADH was made up to give an absorption of 1 at 340 nm in a cuvette of 1 cm path length. The molar absorptivity of NADH is 6220 cm^{-1}. Calculate the concentration of NADH in the solution.

ANSWER

Use the Beer–Lambert law $A = \varepsilon cl$. The equation will need transforming to isolate the term c. This is done by dividing both sides of the equation by εl to give:

$$\frac{A}{\varepsilon l} = \frac{\varepsilon cl}{\varepsilon l}$$

and then cancelling out εl from the right-hand side and substituting in values to give:

$$\frac{1}{6220 \times 1} = c$$

$$1.6 \times 10^{-4} \text{ mol dm}^{-3} = c$$

QUESTION 13.6

A molecule has a molar extinction coefficient of 5400 cm^{-1} at 525 nm. What would be the absorbance of a 10^{-4} mol dm^{-3} solution at 525 nm in a 1 cm path length cuvette?

■ 13.7 THE SPECTROPHOTOMETER

The instrument used to measure light absorbance is called a spectrophotometer. A simplified diagram of a spectrophotometer is shown in Figure 13.12. The way the spectrophotometer works is fairly simple. Light from the light source, which is normally a tungsten lamp for visible light and a deuterium lamp for UV light, falls onto a mirror. The mirror can be rotated to reflect either UV or visible light into the instrument. The selected light is then passed through a slit that can be varied to alter the amount of light that irradiates the sample. A mirror system directs the light to a reflecting prism, which splits the light into its constituent wavelengths. Adjusting the angle of the prism or mirror allows selection of the wavelength of light to be measured. Modern instruments often use other light diffraction methods such as holographic mirrors or diffraction gratings to resolve light into its constituent wavelengths. The light is then reflected through the sample where some is absorbed. Finally the light falls onto a photocell where the light intensity is converted to an electrical signal. The electrical signal produced can be

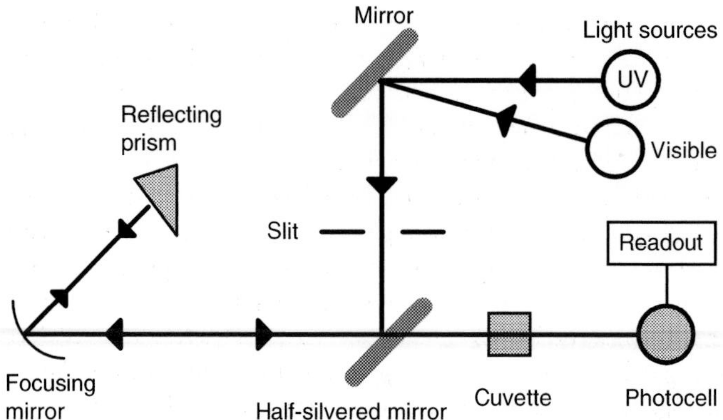

• **Figure 13.12** A simplified diagram of a single-beam UV–visible spectrophotometer.

recorded in a variety of ways such as a dial scale or via an analogue-to-digital converter to a computer.

There are more complicated spectrophotometers which have specialised functions, such as dual-wavelength (measures absorbance changes at two wavelengths simultaneously) or dual-beam instruments (splits the light beam into two and measures the difference between a test and a control beam).

■ 13.8 THE FATE OF ABSORBED LIGHT

The antibonding orbital which an electron occupies as a result of light absorption is unstable and the energy trapped within the bond when light is absorbed is quickly lost. There are several ways in which this can happen (Figure 13.13). These are radiationless loss of energy, resonance energy transfer, chemical reaction and fluorescence.

Radiationless loss of energy involves the electron returning from the antibonding to the bonding orbital without emitting any electromagnetic radiation. This can be due to an overlap between levels in the bonding and antibonding orbitals. From the previous discussion this may not be immediately obvious. An examination of Figure 13.14 may make this easier to understand. Chapters 1 and 2 describe covalent bonds as being flexible,

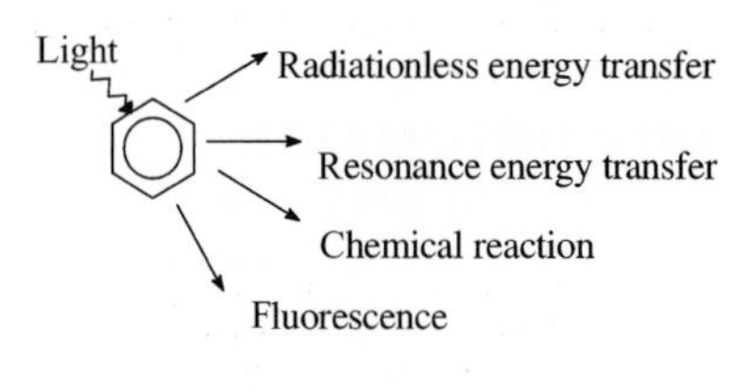

• **Figure 13.13** The fate of absorbed light.

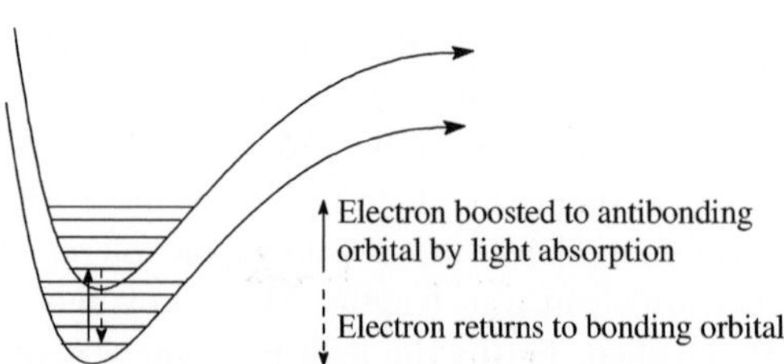

• **Figure 13.14** Radiationless loss of energy due to electrons returning to the ground state via vibrational levels.

with the distance between the atoms joined by the bond varying due to vibration and stretching. There are some stable energies in a bond which are due to different vibrational levels (see right). Light absorption causes a jump to an antibonding orbital. The energy of the antibonding orbital can return via one of the higher vibrational energies of the bonding orbital without emission of radiation.

Bonds showing vibrational energies as thin lines very close to the energy level of the bond

Resonance energy transfer is a process by which absorbed light can be transferred between molecules that are placed close together if there is an overlap between the spectra of the two molecules. The overlap of spectra means that the light (energy) absorption by the first molecule is close enough to the energy required for the second molecule to cause transfer of the absorbed energy, see Figure 13.15. This results in its electron jumping to its antibonding orbital. Resonance energy transfer can only occur if the two molecules are very close together, because the ability of energy to transfer decreases by a factor of 10^6 as the distance between the chromophores is increased. This process is crucial to photosynthesis where the light-harvesting complexes in the thylakoid membrane contain chlorophylls far removed from the reaction centre of the photosystem. The trapped light is then transferred to the reaction centre by resonance energy transfer.

In the photosynthetic reaction centre the energy causes a chemical reaction to occur. The reaction centre contains a pair of chlorophylls so closely aligned that electrons are delocalised over both porphyrin rings. The absorption of energy causes one of the electrons to be transferred from this molecule to form a positively charged chlorophyll dimer. There are lots of other chemical reactions that are driven by light, such as the response to light of photoreceptor cells in the eye, the seasonal changes mediated by phytochrome in plants or the darkening of photographic film.

In certain molecules the structure is such that photon (energy) absorption cannot be lost through radiationless energy and the structure of the molecule is stable. Such molecules can release the energy by the emission of a photon of light. The types of molecules that do this, such as the side chain of the amino acid tryptophan, often contain aromatic rings. The bonding in rigid ring structures must be less able to lose energy by increased vibration. The light emitted by such molecules is inevitably of a higher wavelength than that absorbed and such molecules show two spectra, one for absorption and one for emission. Fluorescence emission is much more difficult to quantify as fluorescence is much more affected by environmental factors than absorption and so the technique is of little analytical use. Treatment of fluorescence emission data is beyond the scope of this text although an easy-to-read introduction to this topic can by found in Freifelder's *Physical Biochemistry*.

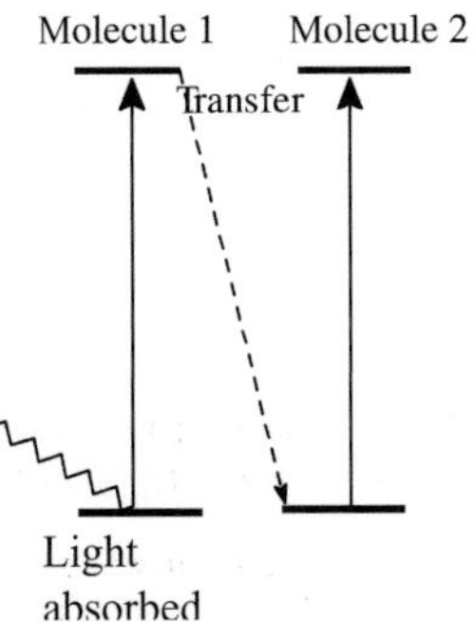

• **Figure 13.15** Transfer of energy by resonance energy transfer.

■ SUMMARY

Light is important to living systems and is part of a greater electromagnetic spectrum. Light can be viewed as both waves and discrete quanta called photons. The energy carried by a photon is inversely related to the wavelength of light. Light energy absorption is dependent on the electronic structure of the molecule with absorption boosting electrons into unoccupied antibonding orbitals. Light energy absorption must relate to the precise energy required to go from bonding to antibonding orbitals. Molecules with conjugated double bonds have a lower energy barrier to the promotion of electrons into antibonding orbitals and such compounds absorb light at higher wavelengths. Light absorption is related to concentration by the Beer–Lambert law, which allows quantitative estimation of concentrations in solution either directly or by colorimetry. Light absorption is measured by using a spectrophotometer. Absorbed light can be lost by radiationless means, resonance energy transfer, chemical reaction or fluorescence.

■ SUGGESTED FURTHER READING

Holme, D.J. and Peck, H. (1998) *Analytical Biochemistry*, 3rd edn, Ch. 2. Longman, Harlow, UK.

Harris, D.A. and Bashford, C.L. (1985) *Spectrophotometry and Spectrofluorimetry – A Practical Approach*. IRL Press, Oxford.

Freifelder, D. (1982) *Physical Biochemistry – Applications to Biochemistry and Molecular Biology*, Chs 14 and 15. W.H. Truman & Co., Oxford.

■ END OF CHAPTER QUESTIONS

Question 13.7 Figure 13.16 shows three compounds. Decide which compound will absorb at the highest and which at the lowest wavelengths.

(a) (b) (c)

• **Figure 13.16** Three molecules containing multiple double bonds.

Question 13.8 A molecule has an absorbance of 0.3 in a 1 cm path length cuvette at 475 nm when dissolved in solution at 0.3 mol dm^{-3}. Calculate the molar extinction coefficient at 475 nm for this molecule.

Question 13.9 Describe the function of the following parts of a single-beam spectrophotometer: (a) reflecting prism, (b) photocell and (c) variable slit.

ANSWERS TO QUESTIONS

CHAPTER 1

Q1.1 (i) Sodium has atomic number 11 (Table 1.4); this is the number of electrons. The number of neutrons is the mass number minus the atomic number: 23 − 11 = 12.

(ii) Symbol, Na; mass number, 23; atomic number, 11, this is written $^{23}_{11}Na$.

Q1.2 $^{35}S \rightarrow {}^{35}Cl$ + β-particle The β-particle can also be written ^{0}e

Q1.3 Full atomic structure   Simplified atomic structure

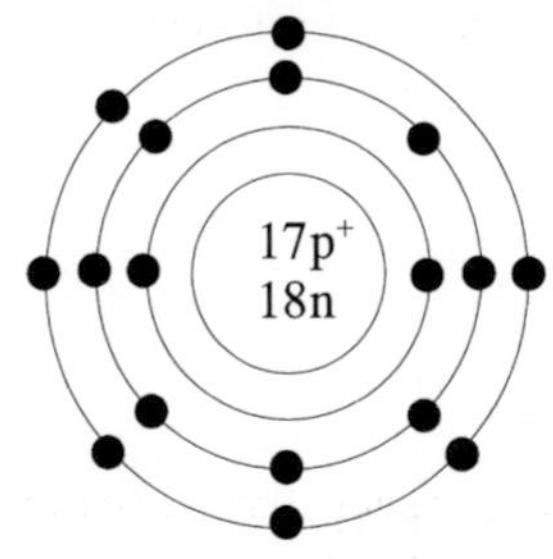

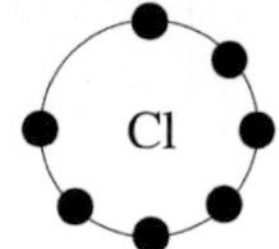

Electron structure: Cl 2.8.7

Q1.4 Electron configuration for oxygen

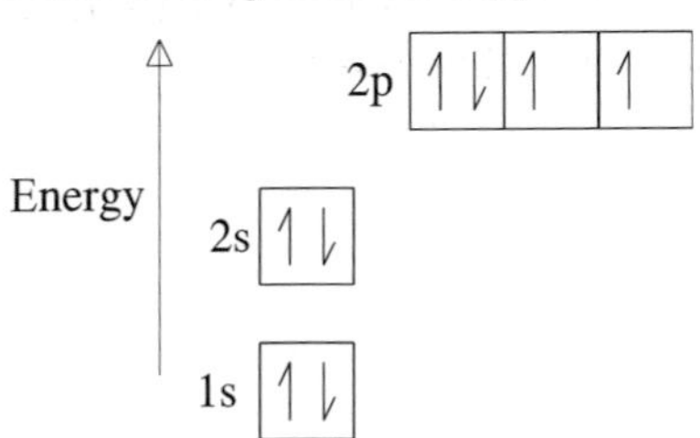

Q1.5 The symbols are B, K, Co, I, Ca and Mo, respectively.

Q1.6

	Rel. at. mass	Atomic no.	Neutrons	Electrons
^{1}H	1	1	0	1
^{2}H	2	1	1	1
^{14}C	14	6	8	6
^{31}P	31	15	16	15
^{37}Cl	37	17	20	17

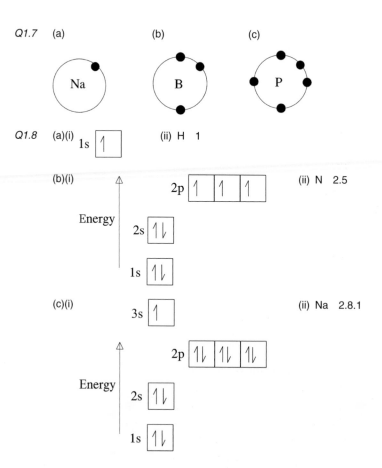

Q1.7 (a) (b) (c)

Na B P

Q1.8 (a)(i) 1s ↿ (ii) H 1

(b)(i) (ii) N 2.5

Energy

2p ↿ | ↿ | ↿

2s ↿⇂

1s ↿⇂

(c)(i) (ii) Na 2.8.1

3s ↿

Energy

2p ↿⇂ | ↿⇂ | ↿⇂

2s ↿⇂

1s ↿⇂

CHAPTER 2

Q2.1

H N H H → H N H H

H—N—H
|
H

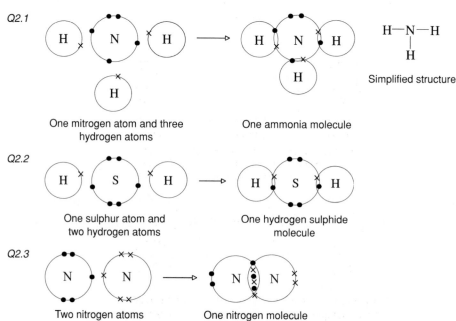

Simplified structure

One mitrogen atom and three hydrogen atoms

One ammonia molecule

Q2.2

H S H → H S H

One sulphur atom and two hydrogen atoms

One hydrogen sulphide molecule

Q2.3

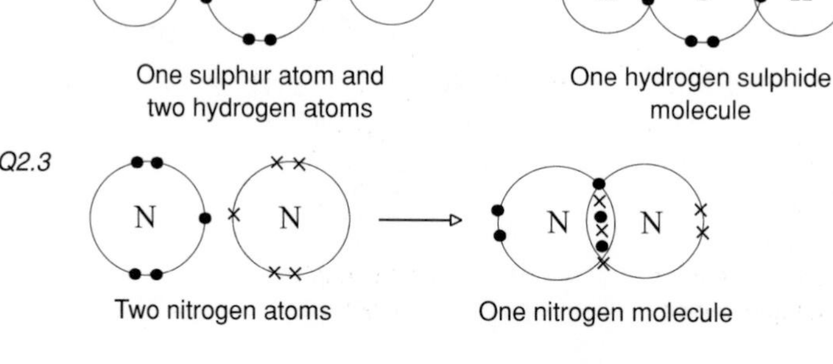

Two nitrogen atoms

One nitrogen molecule

Q2.4

N	H
3	1
N_1	H_3

NH_3

Q2.5

H	S
1	2
H_2	S_1

H_2S

Q2.6

K	SO_4
1	2
K_2	$(SO_4)_1$

K_2SO_4

Q2.7

NH_4	PO_4
1	3
$(NH_4)_3$	$(PO_4)_1$

$(NH_4)_3PO_4$

Q2.8

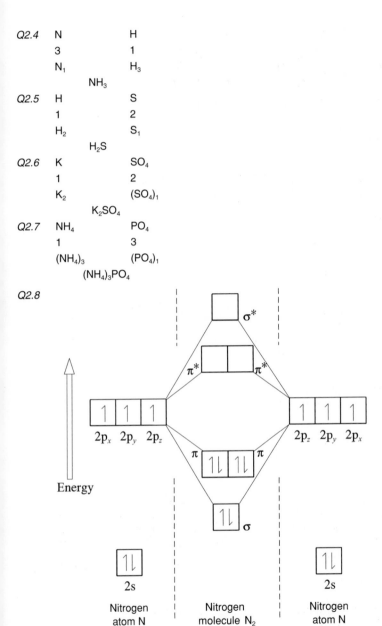

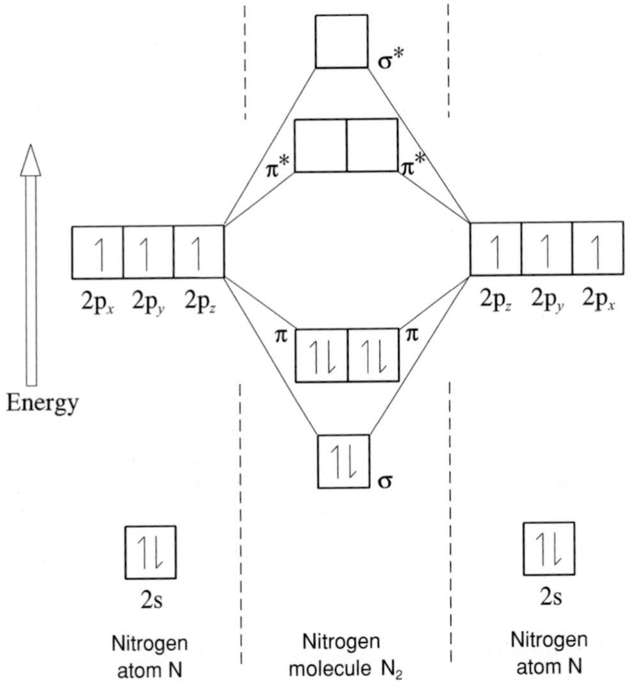

σ*

π* π*

$2p_x$ $2p_y$ $2p_z$ | $2p_z$ $2p_y$ $2p_x$

π π

Energy

σ

2s | 2s

Nitrogen atom N | Nitrogen molecule N_2 | Nitrogen atom N

Molecular orbital energy level diagram to show the formation of the nitrogen molecule N_2 from two nitrogen atoms

Q2.9 (i)

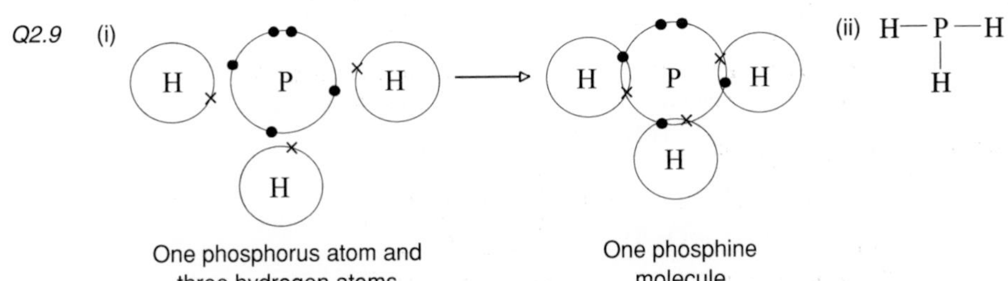

One phosphorus atom and three hydrogen atoms

One phosphine molecule

Q2.10 (i) (ii)

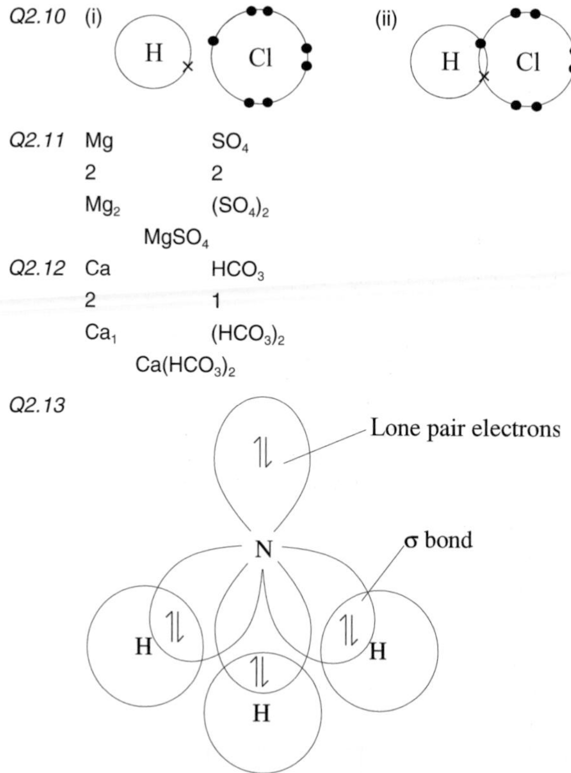

Q2.11

Mg	SO_4
2	2
Mg_2	$(SO_4)_2$

$MgSO_4$

Q2.12

Ca	HCO_3
2	1
Ca_1	$(HCO_3)_2$

$Ca(HCO_3)_2$

Q2.13

CHAPTER 3

Q3.1 Electron configurations: K 2.8.8.1, Cl 2.8.7, K^+ 2.8.8, Cl^- 2.8.8

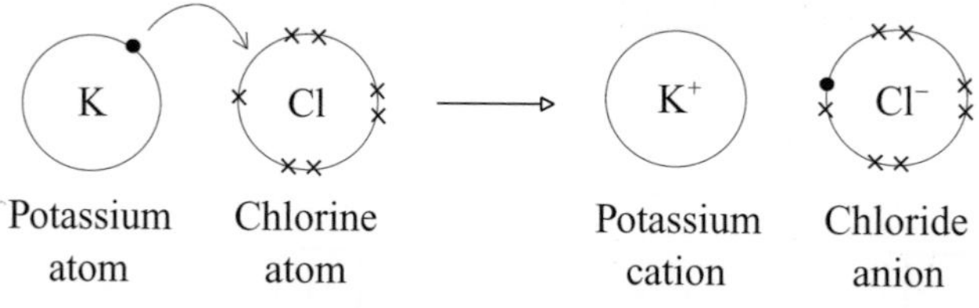

Potassium chloride KCl

Q3.2 (a) From Table 3.2: S = 2.5, O = 3.5, difference is 3.5 − 2.5 = 1.0.

(b) Sulphur less electronegative therefore slightly positive, oxygen more electronegative therefore slightly negative.

Q3.3 $O^{\delta-}{=}C^{\delta+}{=}O^{\delta-} \cdots \overset{\delta-}{O}{=}\overset{\delta+}{C}{=}\overset{\delta-}{O}$

Q3.4 (a)

$H_2N{-}CH_2{-}C({=}O){-}O{-}H$

$H_2N{-}CH_2{-}C({=}O){-}O{-}H$ (the O–H hydrogen hydrogen-bonded (dashed line) to the nitrogen lone pair of the molecule above)

(b)

O (lone pair)
||
$H_2N-CH_2-C-O-H$
⋮
O (lone pair)
||
$H_2N-CH_2-C-O-H$

Q3.5 (a) Electron configurations: Mg 2.8.2, Cl 2.8.7, Mg^{2+} 2.8, 2 × Cl^- 2.8.8

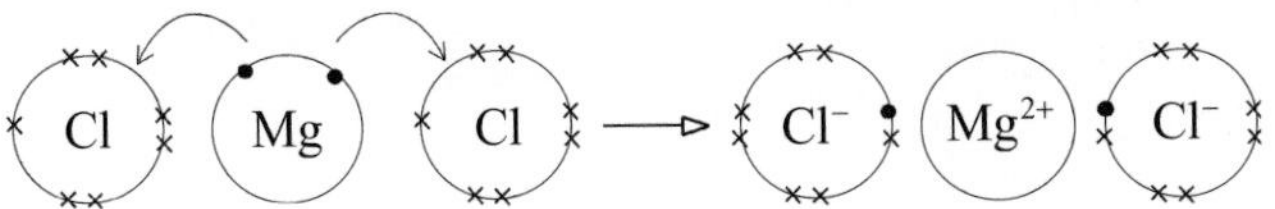

Two chlorine atoms and one magnesium atom

Two chloride anions and one magnesium cation
Magnesium chloride $MgCl_2$

(b)

	Protons	Electrons
Mg	12	12
Mg^{2+}	12	10
Cl	17	17
Cl^-	17	18

Q3.6 (a) From Table 3.2: P = 2.1, C = 2.5, O = 3.5, Cl = 3.0

P—O	3.5 – 2.1 = 1.4
C—O	3.5 – 2.5 = 1.0
C—Cl	3.0 – 2.5 = 0.5

(b) The P—O bond has the largest electronegativity difference and is therefore the most polar.

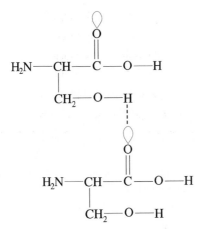

(c) $\overset{\delta+}{C}—\overset{\delta-}{O}$ $\overset{\delta+}{P}—\overset{\delta-}{O}$ $\overset{\delta+}{C}—\overset{\delta-}{Cl}$

Q3.7 (a)

O (lone pair)
||
$H_2N—CH—C—O—H$
|
$CH_2—O—H$
⋮
O (lone pair)
||
$H_2N—CH—C—O—H$
|
$CH_2—O—H$

(b)

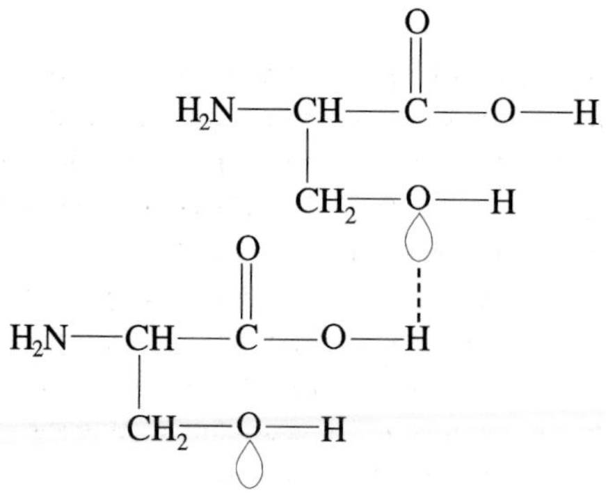

```
                    O
                    ‖
         H2N—CH—C—O—H
               |
              CH2—O—H
                    :
          O         :
          ‖         :
H2N—CH—C—O—H
      |
     CH2—O—H
```

(c)

```
                    O
                    ‖
          H2N—CH—C—O—H
           :     |
           :    CH2—O—H
           O
           ‖
H2N—CH—C—O—H
      |
     CH2—O—H
```

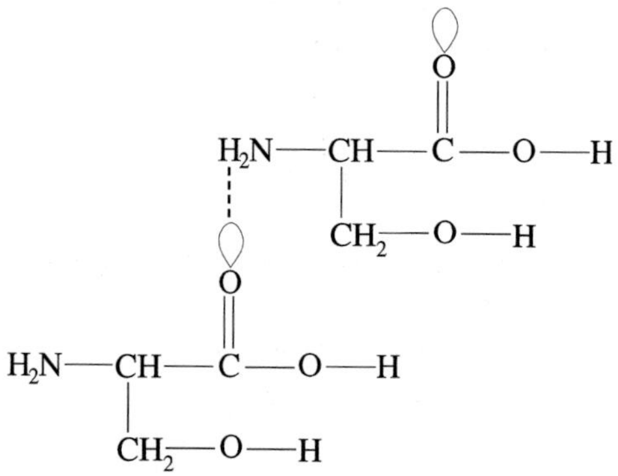

(d)

```
           O
           ‖
H2N—CH—C—O—H
 :     |
 :    CH2—O—H
H2N    O
 |     ‖
 CH—C—O—H
 |
 CH2—O—H
```

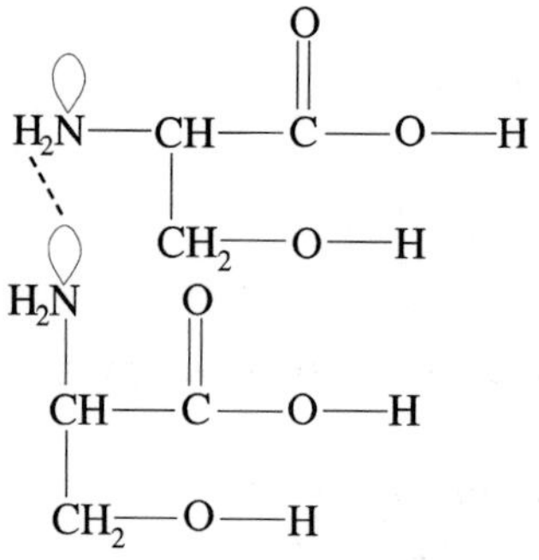

Q3.8 Hydrogen bonding is important in the α-helix and β-pleated sheet structure of proteins (or other examples). The secondary structure of the α-helix is formed by C=O · · · H–N hydrogen bonding between different sections of the same polypeptide chain. In the β-pleated sheet, separate sections of the chain are linked by similar C=O · · · H–N hydrogen bonding.

Q3.9 A protein in aqueous solution has hydrophobic alkyl groups, and these groups preferentially take up positions at the centre of the folded molecule and remote from polar water molecules. This increases the randomness or the entropy of the solution and enhances the stability of the protein (see section 3.7).

Q3.10 (a) Transition metals.

(b) Both electrons come from one of the two atoms making up the bond.

(c) Within the haemoglobin molecule iron is coordinated by five nitrogen ligands: four from the porphyrin ring and one from a histidine group. The iron is held tightly in a hydrophobic environment and is able to receive an oxygen molecule as a ligand and bind it loosely. This process is significant in carrying oxygen for respiration.

CHAPTER 4

Q4.1 (a) Increasing the concentration of CoASH threefold while holding the concentration of ethanoyl chloride constant increases the rate of the reaction threefold (from 3.4×10^{-3} mmol dm^{-3} min^{-1} to 1.02×10^{-2} mmol dm^{-3} min^{-1}). The rate is therefore directly proportional to the concentration of CoASH, i.e. the rate varies according to $[\text{CoASH}]^1$. The reaction is therefore first order with respect to CoASH.

Doubling the concentration of ethanoyl chloride while holding the concentration of CoASH constant doubles the rate of the reaction (from 3.4×10^{-3} mmol dm^{-3} min^{-1} to 6.8×10^{-3} mmol dm^{-3} min^{-1}. The rate is therefore directly proportional to the concentration of ethanoyl chloride, i.e. the rate varies according to $[\text{ethanoyl chloride}]^1$. The reaction is therefore first order with respect to ethanoyl chloride.

(b) Since 1 + 1 = 2, the reaction is second order overall.

Q4.2 (a) The expression for the equilibrium constant is:

$$K_{eq} = \frac{[\text{G6P}]}{[\text{G1P}]}$$

(b) Substituting the given concentration values in this expression gives:

$$K_{eq} = \frac{5.8 \times 10^{-2} \text{ mol dm}^{-3}}{3.0 \times 10^{-3} \text{ mol dm}^{-3}}$$

$$= 19.3$$

Q4.3 (a) The expression for the equilibrium constant is:

$$K_{eq} = \frac{k_f}{k_b}$$

(b) Substituting the given rate constant values in this expression gives:

$$K_{eq} = \frac{1.3 \times 10^{7} \text{ mol}^{-1} \text{ dm}^{3} \text{ s}^{-1}}{3.2 \text{ s}^{-1}}$$

$$= 4.06 \times 10^{6} \text{ mol}^{-1} \text{ dm}^{3}$$

Q4.4 (a) Doubling the concentration of B while holding the concentration of A constant increases the rate of reaction fourfold (from 2.86×10^{-5} mmol dm^{-3} min^{-1} to 1.14×10^{-4} mmol dm^{-3} min^{-1}). The rate therefore increases according to $[\text{B}]^2$. The reaction is therefore second order with respect to B.

Doubling the concentration of A while holding the concentration of B constant has no effect on the rate of the reaction. The reaction is therefore zero order with respect to A.

(b) Since 2 + 0 = 2, the reaction is second order overall.

Q4.5 (a) The expression for the equilibrium constant is:

$$K_{eq} = \frac{[\text{C}_5\text{H}_{11}\text{O}_5\text{COO}^-][\text{H}^+]}{[\text{C}_5\text{H}_{11}\text{O}_5\text{COOH}]}$$

(b) Substituting the given concentration values gives:

$$K_{eq} = \frac{6.6 \times 10^{-4}\ \text{mol dm}^{-3} \times 6.6 \times 10^{-4}\ \text{mol dm}^{-3}}{3.2 \times 10^{-2}\ \text{mol dm}^{-3}}$$

$$= 1.36 \times 10^{-5}\ \text{mol dm}^{-3}$$

Q4.6 The required expression for K_{eq} is:

$$K_{eq} = \frac{k_f}{k_b}$$

and substituting the given rate constant values gives:

$$K_{eq} = \frac{7.2 \times 10^{-3}\ s^{-1}}{2.2 \times 10^{-5}\ s^{-1}}$$

$$= 327.3$$

CHAPTER 5

Q5.1 (a) Nitric acid is HNO_3 and the required equation is:

$$HNO_3 + H_2O \rightarrow H_3O^+ + NO_3^-$$

nitric acid + water → hydronium ion + nitrate ion

(b) Potassium hydroxide is KOH and the required equation is:

$$KOH \rightarrow K^+ + OH^-$$

potassium hydroxide → potassium ion + hydroxide ion

Q5.2 For a strong acid:

$$pH = -\log_{10} C$$

Substituting the given concentration value gives:

$$pH = -\log_{10} 0.1$$
$$= -(-1)$$
$$= 1$$

pH of a 0.1 mol dm^{-3} solution of HCl = 1

Q5.3 For a strong base:

$$pH = pK_w + \log_{10} C$$

Substituting in the values of pK_w and concentration gives:

$$pH = 14 + \log_{10} 0.015$$
$$= 14 + (-1.82)$$
$$= 14 - 1.82$$
$$= 12.18$$

pH of a 0.015 mol dm^{-3} solution of NaOH = 12.18

Q5.4 For a weak acid:

$$pH = \tfrac{1}{2}pK_a - \tfrac{1}{2}\log_{10} C$$

Substituting in the given values for pK_a and concentration gives:

$$\begin{aligned} pH &= \tfrac{1}{2}(3.86) - \tfrac{1}{2}(\log_{10} 0.025) \\ &= 1.93 - \tfrac{1}{2}(-1.60) \\ &= 1.93 + 0.80 \\ &= 2.73 \end{aligned}$$

pH of a 0.025 mol dm^{-3} solution of lactic acid = 2.73

Q5.5 For a weak base when pK_b is given,

$$pH = pK_w - \tfrac{1}{2}pK_b + \tfrac{1}{2}\log_{10} C$$

Substituting in the values for pK_w, pK_b and concentration gives:

$$\begin{aligned} pH &= 14 - \tfrac{1}{2}(3.34) + \tfrac{1}{2}(\log_{10} 0.005) \\ &= 14 - 1.67 + \tfrac{1}{2}(-2.30) \\ &= 14 - 1.67 - 1.15 \\ &= 11.18 \end{aligned}$$

pH of a 0.005 mol dm^{-3} solution of methylamine = 11.18

Q5.6 For a weak base when pK_a is given:

$$pH = \tfrac{1}{2}pK_w + \tfrac{1}{2}pK_a + \tfrac{1}{2}\log_{10} C$$

Substituting in the values for pK_w, pK_a and concentration gives:

$$\begin{aligned} pH &= \tfrac{1}{2}(14) + \tfrac{1}{2}(9.81) + \tfrac{1}{2}(\log_{10} 0.015) \\ &= 7 + 4.91 + \tfrac{1}{2}(-1.82) \\ &= 7 + 4.91 - 0.91 \\ &= 11.00 \end{aligned}$$

pH of a 0.015 mol dm^{-3} solution of trimethylamine = 11.00

Q5.7 For a salt of a strong base with a weak acid:

$$pH = \tfrac{1}{2}pK_w + \tfrac{1}{2}pK_a + \tfrac{1}{2}\log_{10} C$$

Substituting in the values of pK_w, pK_a and concentration gives:

$$\begin{aligned} pH &= \tfrac{1}{2}(14) + \tfrac{1}{2}(3.75) + \tfrac{1}{2}(\log_{10} 0.02) \\ &= 7 + 1.88 + \tfrac{1}{2}(-1.70) \\ &= 7 + 1.88 - 0.85 \\ &= 8.03 \end{aligned}$$

pH of a 0.02 mol dm^{-3} solution of sodium methanoate = 8.03

Q5.8 For a salt of a weak base with a strong acid:

$$pH = \tfrac{1}{2}pK_a - \tfrac{1}{2}\log_{10} C$$

Substituting in the given values of pK_a and concentration gives:

$$\begin{aligned} pH &= \tfrac{1}{2}(10.99) - \tfrac{1}{2}(\log_{10} 0.01) \\ &= 5.50 - \tfrac{1}{2}(-2) \\ &= 5.50 + 1 \\ &= 6.50 \end{aligned}$$

pH of a 0.01 mol dm^{-3} solution of diethylammonium chloride = 6.50

Q5.9 In mixing the two solutions, both have been diluted. Volume of final solution = 500 cm^3

Concentration of ethanoic acid in this solution = $\dfrac{200}{500} \times 0.2$ mol dm^{-3} = 0.08 mol dm^{-3}

Concentration of sodium ethanoate in this solution = $\dfrac{300}{500} \times 0.15$ mol dm^{-3} = 0.09 mol dm^{-3}

We then use the Henderson–Hasselbach equation:

$$pH = pK_a + \log_{10}\frac{[\text{unprotonated species}]}{[\text{protonated species}]}$$

In this case the unprotonated species is the ethanoate anion, the concentration of which is equal to the concentration of the sodium ethanoate. The protonated species is the ethanoic acid, the concentration of which is equal to the concentration of the acid in the solution. Substituting in the values calculated above and the given value of pK_a gives:

$$pH = 4.75 + \log_{10}\frac{0.09}{0.08}$$

$$\begin{aligned} &= 4.75 + \log_{10} 1.125 \\ &= 4.75 + 0.05 \\ &= 4.8 \end{aligned}$$

pH of the buffer = 4.8

Q5.10 The first step is to calculate the ratio of salt to base required. To do this, we use the Henderson–Hasselbach equation:

$$pH = pK_a + \log_{10}\frac{[\text{unprotonated species}]}{[\text{protonated species}]}$$

Substituting in the desired value of pH and the given value of pK_a gives:

$$8.0 = 8.08 + \log_{10}\frac{[\text{unprotonated species}]}{[\text{protonated species}]}$$

$$\log_{10}\frac{[\text{unprotonated species}]}{[\text{protonated species}]} = 8.0 - 8.08$$

$$= -0.08$$

$$\frac{[\text{unprotonated species}]}{[\text{protonated species}]} = 0.832$$

Let the volume of 0.5 mol dm^{-3} Tris = V cm^3

Therefore, volume of 1.0 mol dm^{-3} acid required = 1000 – V cm^3

Concentration of Tris after dilution to 1000 cm^3 will be $\frac{0.5 \times V}{1000}$ mol dm^{-3}

Concentration of acid after dilution to 1000 cm^3 will be $\frac{1.0 \times (1000 - V)}{1000}$ mol dm^{-3}

In this case, Tris is the unprotonated species and the acid the protonated species. We therefore have:

$$\frac{0.5 \times V}{1000} \div \frac{1.0 \times (1000 - V)}{1000} = 0.832$$

$$\frac{0.5 \times V}{1000} \times \frac{1000}{1.0 \times (1000 - V)} = 0.832$$

$$\frac{500 \times V}{1000 \times (1000 - V)} = 0.832$$

$$500 \times V = 832 \times (1000 - V)$$
$$= 832\,000 - 832\ V$$

$$1332\ V = 832\,000$$

$$V = \frac{832\,000}{1332}\ \text{cm}^3$$

$$= 624.6\ \text{cm}^3$$

Therefore, in order to prepare the required amount of buffer at pH 8, 624.6 cm^3 of the 0.5 mol dm^{-3} Tris must be mixed with 375.4 cm^3 of the 1.0 mol dm^{-3} acid.

Q5.11 Number of moles of sodium hydroxide in 18.70 cm^3 0.1000 mol dm^{-3} solution $= \frac{0.1000}{1000} \times 18.70 = 0.00187$

From the reaction equation, this number of moles will react with an equal number of moles of ethanoic acid.

Number of moles of ethanoic acid = 0.00187

This is present in 25.00 cm^3

1 litre must therefore contain $\frac{0.00187}{25.00} \times 1000$ moles = 0.0748 moles

Concentration of the ethanoic acid = 0.0748 mol dm^{-3}

Q5.12 Number of moles of sulphuric acid in 15.25 cm^3 of 0.05 mol dm^{-3} solution $= \frac{0.05}{1000} \times 15.25$

= 0.0007625

From the reaction equation, this number of moles of sulphuric acid will react with twice as many moles of sodium hydroxide.

Number of moles of sodium hydroxide = 2 × 0.0007625 = 0.001525

This is present in 25.00 cm^3

1 litre must therefore contain $\frac{0.001525}{25.00} \times 1000$ moles = 0.061 moles

Concentration of the sodium hydroxide = 0.061 mol dm^{-3}

Q5.13 (a) Methanoic acid reacts with water as follows:

$$HCOOH + H_2O \rightleftharpoons H_3O^+ + HCOO^-$$

methanoic acid + water ⇌ hydronium ion + methanoate ion

(b) Trimethylamine reacts with water as follows:

$$N(CH_3)_3 + H_2O \rightleftharpoons HN(CH_3)_3^+ + OH^-$$

trimethylamine + water ⇌ trimethylammonium ion + hydroxide ion

Q5.14 (a) For a strong acid:

$$pH = -\log_{10} C$$

Substituting in the given concentration value gives:

$$\begin{aligned} pH &= -\log_{10} 0.15 \\ &= -(-0.82) \\ &= 0.82 \end{aligned}$$

pH of 0.15 mol dm^{-3} hydrochloric acid = 0.82

(b) For a strong base:

$$pH = pK_w + \log_{10} C$$

Substituting in the values of pK_w and concentration gives:

$$\begin{aligned} pH &= 14 + \log_{10} 0.01 \\ &= 14 + (-2) \\ &= 14 - 2 \\ &= 12 \end{aligned}$$

pH of 0.01 mol dm^{-3} potassium hydroxide = 12

Q5.15 (a) For a weak acid:

$$pH = \tfrac{1}{2}pK_a - \tfrac{1}{2}\log_{10} C$$

Substituting in the given values of pK_a and concentration gives:

$$\begin{aligned} pH &= \tfrac{1}{2}(9.89) - \tfrac{1}{2}(\log_{10} 0.05) \\ &= 4.95 - \tfrac{1}{2}(-1.30) \\ &= 4.95 + 0.65 \\ &= 5.6 \end{aligned}$$

pH of a 0.05 mol dm^{-3} solution of phenol = 5.6

(b) For a weak base for which pK_a is given:

$$pH = \tfrac{1}{2}pK_w + \tfrac{1}{2}pK_a + \tfrac{1}{2}\log_{10} C$$

Substituting in the given values of pK_w, pK_a and concentration gives:

$$\begin{aligned} pH &= \tfrac{1}{2}(14) + \tfrac{1}{2}(10.76) + \tfrac{1}{2}\log_{10} 0.1 \\ &= 7 + 5.38 + \tfrac{1}{2}(-1.0) \\ &= 7 + 5.38 - 0.50 \\ &= 11.88 \end{aligned}$$

pH of a 0.1 mol dm^{-3} solution of triethylamine = 11.88

Q5.16 (a) For a salt of a strong base with a weak acid:

$$pH = \tfrac{1}{2}pK_w + \tfrac{1}{2}pK_a + \tfrac{1}{2}\log_{10} C$$

Substituting in the values of pK_w, pK_a and concentration gives:

$$\begin{aligned} pH &= \tfrac{1}{2}(14) + \tfrac{1}{2}(4.87) + \tfrac{1}{2}\log_{10} 0.25 \\ &= 7 + 2.44 + \tfrac{1}{2}(-0.60) \\ &= 7 + 2.44 - 0.30 \\ &= 9.14 \end{aligned}$$

pH of a 0.25 mol dm^{-3} solution of sodium propionate = 8.84

(b) For a salt of a weak base with a strong acid:

$$pH = \tfrac{1}{2}pK_a - \tfrac{1}{2}\log_{10} C$$

Substituting in the given values of pK_a and concentration gives:

$$\begin{aligned} pH &= \tfrac{1}{2}(10.66) - \tfrac{1}{2}(\log_{10} 0.025) \\ &= 5.33 - \tfrac{1}{2}(-1.60) \\ &= 5.33 + 0.80 \\ &= 6.13 \end{aligned}$$

pH of a 0.025 mol dm^{-3} solution of methylammonium chloride = 6.13

Q5.17 (a) In this case, solid sodium ethanoate is being added to a solution of ethanoic acid, so no dilution occurs. Ethanoic acid is the protonated species and ethanoate ion the unprotonated species.

Using the Henderson–Hasselbach equation:

$$pH = pK_a + \log_{10}\frac{[\text{unprotonated species}]}{[\text{protonated species}]}$$

and substituting in the required pH and the given values of pK_a and concentration of ethanoic acid gives:

$$4.5 = 4.75 + \log_{10}\frac{[\text{ethanoate ion}]}{0.2}$$

$$\log_{10}\frac{[\text{ethanoate ion}]}{0.2} = 4.5 - 4.75 = -0.25$$

$$\frac{[\text{ethanoate ion}]}{0.2} = 0.562$$

$$[\text{ethanoate ion}] = 0.562 \times 0.2$$
$$= 0.112 \text{ mol dm}^{-3}$$

However, we require only 100 cm^3 rather than 1 dm^3, so:

Number of moles of sodium ethanoate required $= \dfrac{0.112}{10} = 0.0112$

(b) Formula of sodium ethanoate is CH_3COONa.
Its molar mass is the sum of the masses of its constituent atoms.

$$\text{Molar mass} = 12 + 3 \times 1 + 12 + 2 \times 16 + 23$$
$$= 82 \text{ g mol}^{-1}$$

Therefore we require 0.0112 × 82 g = 0.9184 g

Q5.18 Let the volume of ammonia solution required be V cm^3
Then, the volume of hydrochloric acid required = $500 - V$ cm^3
Both solutions will be diluted on mixing.
On mixing, the hydrochloric acid reacts with the ammonia to form ammonium chloride according to the equation:

$$NH_3 + HCl \rightarrow NH_4Cl$$

Each mole of HCl added produces one mole of ammonium chloride, so that:
Concentration of ammonium chloride in final solution = concentration of acid added
Ammonia is the unprotonated species and ammonium ion the protonated species.

Concentration of ammonia in mixed solution $= \dfrac{0.5\,V}{500}$

Concentration of ammonium chloride in mixed solution $= \dfrac{2.0(500 - V)}{500}$

Using the Henderson–Hasselbach equation:

$$\text{pH} = \text{p}K_a + \log_{10}\frac{[\text{unprotonated species}]}{[\text{protonated species}]}$$

and substituting in the required pH, the given pK_a and the expressions for the concentrations gives:

$$7.8 = 9.25 + \log_{10}\frac{[\text{unprotonated species}]}{[\text{protonated species}]}$$

$$\log_{10}\frac{[\text{unprotonated species}]}{[\text{protonated species}]} = 7.8 - 9.25$$

$$= -1.45$$

$$\frac{[\text{unprotonated species}]}{[\text{protonated species}]} = 0.0355$$

$$\frac{0.5\,V}{500} \div \frac{2.0(500 - V)}{500} = 0.0355$$

$$\frac{0.5\,V}{500} \times \frac{500}{2.0(500 - V)} = 0.0355$$

$$\frac{250\,V}{1000\,(500 - V)} = 0.0355$$

$$250\ V = 35.5(500 - V)$$
$$= 17\,750 - 35.5\ V$$

$$285.5\ V = 17\,750$$

$$V = 62.2\ \text{cm}^3$$

Therefore, 62.2 cm^3 of the ammonia solution must be mixed with 437.8 cm^3 of the hydrochloric acid solution.

Q5.19 The reaction equation is:

$$CH_3COOH + NaOH \rightarrow CH_3COONa + H_2O$$

Number of moles of sodium hydroxide used = $\frac{0.100}{1000} \times 19.6 = 0.00196$

From the reaction equation, this will react with an equal number of moles of ethanoic acid.
Number of moles of ethanoic acid present = 0.00196
This is present in 25.00 cm^3 of solution.

1 dm^3 would therefore contain = $\frac{0.00196}{25.00} \times 1000$ moles = 0.0784 moles

Concentration of ethanoic acid = 0.0784 mol dm^{-3}.

CHAPTER 6

Q6.1

carbon dioxide + water ⇌ carbonic acid

$$CO_2 + H_2O \rightleftharpoons H_2CO_3$$

carbonic acid ⇌ hydrogen ions + hydrogencarbonate ions

$$H_2CO_3 \rightleftharpoons H^+ + HCO_3^-$$

hydrogencarbonate ions ⇌ hydrogen ions + carbonate ions

$$HCO_3^- \rightleftharpoons H^+ + CO_3^{2-}$$

Calcium is mobilised when hydrogen ions in rainwater containing dissolved carbon dioxide bring calcium carbonate into solution as calcium hydrogencarbonate. Carbon is mobilised as hydrogencarbonate in the same reaction.

calcium carbonate + carbonic acid ⇌ calcium hydrogencarbonate

$$CaCO_3 + H_2CO_3 \rightleftharpoons Ca(HCO_3)_2$$

Q6.2 Respiration in animals and plants releases carbon dioxide and water by the cellular breakdown of sugars.
Combustion in air of fossil fuels such as coal or oil gives carbon dioxide.
Microbiological decay of plants and animals breaks down carbon-containing biomolecules to give ultimately carbon dioxide and water.

Q6.3 Molecular formulae (a) C_5H_{12} (b) C_6H_{14}
Brief structural formulae (a) $CH_3CH_2CH_2CH_2CH_3$ (b) $CH_3CH_2CH_2CH_2CH_2CH_3$

Q6.4 (a) C_2H_6, ethane gives $CH_3CH_2–$, the ethyl group.

(b) C_4H_{10}, butane gives $CH_3CH_2CH_2CH_2–$, the butyl group.

Q6.5

$$\text{(a)}\quad \begin{array}{ccccccc} & & & H & & & \\ & & & | & & & \\ & H & & O & & H & \\ & | & & | & & | & \\ H- & C & - & C & - & C & -H \\ & | & & | & & | & \\ & H & & H & & H & \end{array} \qquad \text{(b)}\quad \begin{array}{ccccccccc} & & & & & H & & & \\ & & & & & | & & & \\ & H & & H & & O & & H & \\ & | & & | & & | & & | & \\ H- & C & - & C & - & C & - & C & -H \\ & | & & | & & | & & | & \\ & H & & H & & H & & H & \end{array}$$

It seems reasonable to consider the primary alcohols (a) $CH_3CH_2CH_2OH$ and (b) $CH_3CH_2CH_2CH_2OH$ as products in this reaction. However the nature of the reaction intermediates gives the secondary alcohols. The detailed mechanism is discussed in GCSE 'A' level chemistry texts.

Q6.6

$$\text{(a)}\quad \begin{array}{ccccccc} & H & & H & & H & \\ & | & & | & & | & \\ H- & C & - & C & - & C & -O-H \\ & | & & | & & | & \\ & H & & H & & H & \end{array}$$

$$\text{(b)}\quad \begin{array}{ccccccccccc} & H & & H & & H & & H & & H & \\ & | & & | & & | & & | & & | & \\ H- & C & - & C & - & C & - & C & - & C & -O-H \\ & | & & | & & | & & | & & | & \\ & H & & H & & H & & H & & H & \end{array}$$

$$\text{(c)}\quad \begin{array}{ccccccccc} & H & & H & & H & & H & \\ & | & & | & & | & & | & \\ H- & C & - & C & - & C & - & C & -O-H \\ & | & & | & & | & & | & \\ & H & & H & & H & & H & \end{array}$$

Structural formulae for secondary, rather than primary, alcohols are equally acceptable as answers.

Q6.7 (a) C_2H_6O is CH_3CH_2OH, ethanol.

(b) $C_4H_{10}O$ is $CH_3CH_2CH_2CH_2OH$, butanol.

(c) CH_4O is CH_3OH, methanol.

Q6.8 (a) Carbon dioxide and water.

(b) Ethanoic acid (acetic acid) or carbon dioxide and water.

(c) Propene.

Q6.9

$$\begin{array}{l} H_3N^+\underset{\displaystyle \overset{|}{COO^-}}{C}HCH_2SH \end{array}$$

Q6.10 (a), (c) and (d) are aldehydes; (b), (e) and (f) are ketones.

Q6.11 (a) Propanoic acid.

(b) Methanoic acid (formic acid).

(c) Ethane dioic acid (oxalic acid).

Q6.12 $CH_3CH_2CH_2COOH \rightleftharpoons CH_3CH_2CH_2COO^- + H^+$

Q6.13

(a) $CH_3-C(=O)-O-H$, with the hydroxyl O lone pair $\cdots H-O-H$

(b) $CH_3-C(=O)-O-H \cdots O(H)-H$

Q6.14 (a), (c) and (e) are primary amines; (b) and (f) are secondary amines; (d) is a tertiary amine.

Q6.15 (a) $CH_3CH_2NH_3^+$ (b) $(CH_3)_3NH^+$
(c) $HOCH_2CH_2NH_3^+$ (d) $CH_3NH_2^+CH_3$

Q6.16 (a) Coal is formed when dead plant material becomes buried and subjected to heat and pressure in the earth's crust over very long time periods.
(b) The calcareous shells of tiny marine invertebrates fall in vast numbers to the bed of shallow seas when the organisms die. The layers of sediment formed are heated and crushed in the earth.

Q6.17 (a) Calcium carbonate is solubilised by slightly acid rainwater and taken up by the roots of plants.
(b) Carbon dioxide combines with water in the green leaves of plants during photosynthesis in sunlight. Carbohydrates are formed and become the basis of metabolic processes in plants. Subsequently animals eat the plants.

Q6.18 (a) CH_3CH_2OH (b) $CH_3CH_2CH_2CH_2OH$ (or isomer)

Q6.19 (a) $H_2N—\underset{\displaystyle COOH}{\underset{|}{CH}}—CH_2—S—S—CH_2—\underset{\displaystyle COOH}{\underset{|}{CH}}—NH_2$

(b) The –S–S– covalent linkage can be formed between cysteine groups in different parts of a polypeptide chain. This can be a significant factor in the stabilisation of the polypeptide structure. The enzyme ribonuclease (section 6.7) is an important example.

Q6.20 (a) Aldehydes are formed from (i) and (iii).
(b) (i) Ethanal, (ii) Butanone, (iii) Butanal.
(c) (i) CH_3CHO (ii) $CH_3CH_2COCH_3$ (iii) $CH_3CH_2CH_2CHO$

Q6.21 (a) Ethane, CH_3CH_3 and carbon dioxide.
(b) Propanal, CH_3CH_2CHO and carbon dioxide.

Q6.22

(a) 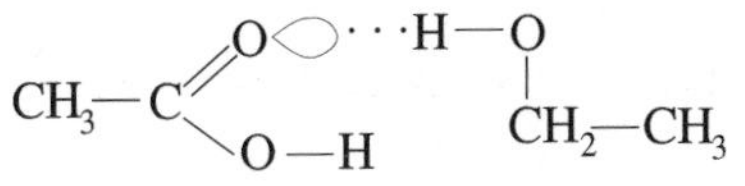 $CH_3—C(=O\cdots H—O—CH_2—CH_3)—O—H$

(b) 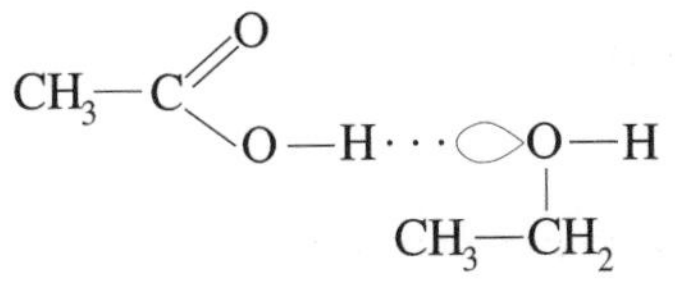 $CH_3—C(=O)—O—H\cdots O(—H)—CH_2—CH_3$

Q6.23 methylamine + hydrochloric acid → methylammonium chloride

$$CH_3NH_2 + HCl \rightarrow CH_3NH_3^+Cl^-$$

CHAPTER 7

Q7.1 The prefix 'hexadeca-' means that the carboxylic acid contains 16 carbons. The 'dien' indicates the presence of two carbon–carbon double bonds, the *cis* indicating the configuration whilst the numbers 9,12 identify the positions of the double bonds as being between carbons 9 and 10 and 12 and 13, respectively. The structure can now be drawn as:

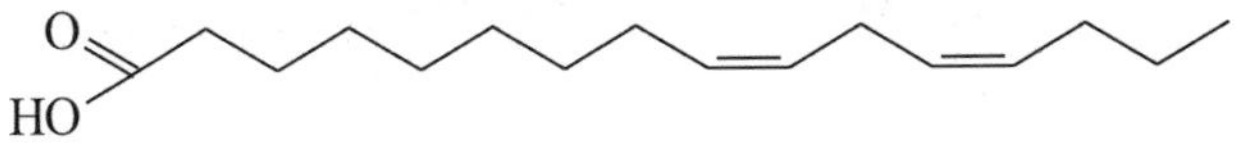

O
HO

Q7.2 First count the number of carbons in each carboxylic acid. In Figure 7.3(b) both carboxylic acids are 12 carbons long. The presence of a *cis* double bond in (i) means that the alkyl chain will be bent and this will therefore decrease van der Waals interactions, lowering the melting temperature. Acid (ii) therefore has a higher melting temperature than (i).

Q7.3 Draw methanoic acid (HCOOH) and ethanol ($HOCH_2CH_3$). Remove the –OH group from the acid (HCO–) and the hydroxyl from the alcohol ($–OCH_2CH_3$). Join up the two bonds to give:

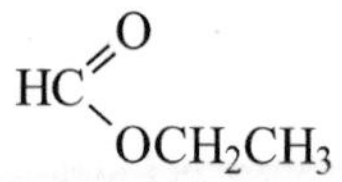

$$HC(=O)OCH_2CH_3$$

The name is derived by referring to the acid by its ion form, methanoate, and the alcohol, ethanol, as an alkyl group, in this case ethyl. The name is therefore ethyl methanoate.

Q7.4 The presence of *cis* double bonds in glyceryl trioleate means that the alkyl chain will be bent. This will therefore decrease van der Waals interactions, lowering the melting point to below room temperature and making this compound an oil. There are no double bonds in glyceryl tristearate and the melting point is above room temperature making this compound a fat.

Q7.5 Count the carbons. Sugar (i) has five carbons and is a pentose, sugar (ii) has three carbons and is a triose. Find the carbonyl group (C=O). Sugar (i) has the carbonyl group in the middle of the chain and is a ketose whilst sugar (ii) has the carbonyl at the end of the chain and is an aldose. Sugar (i) is a ketopentose and sugar (ii) is an aldotriose.

Q7.6 Find the carbonyl group. The sugar is then numbered to give the carbonyl carbon the lowest number. In sugar (i) and (ii) the numbering is from the top of the sugar.

Q7.7 (a) Carbons at the end of a chain are not chiral as they have two hydrogen atoms attached. The carbonyl carbon is not chiral. The carbons shown with an asterisk are chiral.

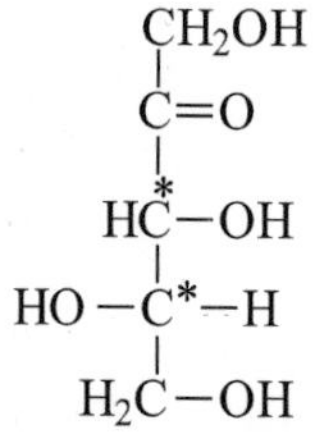

$$\begin{array}{c} CH_2OH \\ | \\ C{=}O \\ | \\ HC^{*}{-}OH \\ | \\ HO{-}C^{*}{-}H \\ | \\ H_2C{-}OH \end{array}$$

(b) The number of stereoisomers is 2^n where n is the number of chiral centres. Thus there are 2^2 or 4 stereoisomers for this ketopentose.

Q7.8 Find the anomeric carbon. This is the one next to the ring oxygen that also contains a hydroxyl group. This is shown in bold:

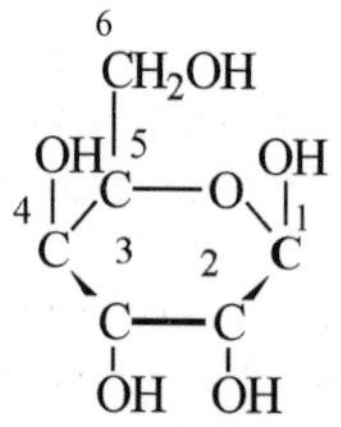

6 CH_2OH on C5; ring C5–O–C1–C2–C3–C4; OH on C4, C1, C2 and C3 (numbers 1–6 as labelled)

This carbon is given the lowest possible number. In this case it is carbon 1 and other carbons are numbered clockwise from this as shown.

Q7.9 Identify the anomeric carbons as described for the answer to Question 7.8. The sugar is reducing if one of the anomeric carbons is attached to a hydroxyl group. This is not the case in this diagram.

Q7.10 Count the number of carbons; in this case 20. The parent acid is eicosanoic acid. Count the number of double bonds; in this case two. The parent alkene acid is eicosadienoic acid. Identify the position and orientation of the double bonds. They are both *cis* double bonds and are at carbons 6 and 9. This makes the full name 'all-*cis*-6,9-eicosadienoic acid'.

Q7.11 Draw ethanoic acid (CH_3COOH) and methanol ($HOCH_3$). Remove the —OH group from the acid (CH_3CO—) and the hydroxyl from the alcohol (—OCH_3). Join up the two bonds to give

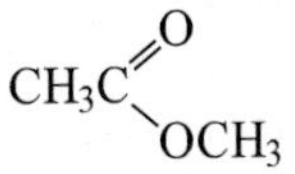

$$CH_3C(=O)OCH_3$$

The name is derived by referring to the acid by its ion form, ethanoate, and the alcohol, methanol, as an alkyl group, in this case methyl. The name is therefore methyl ethanoate.

Q7.12 (a) The carbonyl carbon in a sugar can form a bond with a hydroxyl resulting in a hemiacetal or hemiketal. The carbonyl carbon is not chiral in the open chain form but becomes chiral when a cyclic sugar is formed. Two possible isomers are formed when sugars form furan or pyran ring structures. These isomers are called anomers. (b) α-anomers have the anomeric carbon hydroxyls in the 'down' position whilst β-anomers are in the 'up' position. Identify the anomeric carbon as described in the answer to Question 7.8. Sugars (ii) and (iii) are α-anomers and sugars (i) and (iv) are β-anomers.

Q7.13 (a)

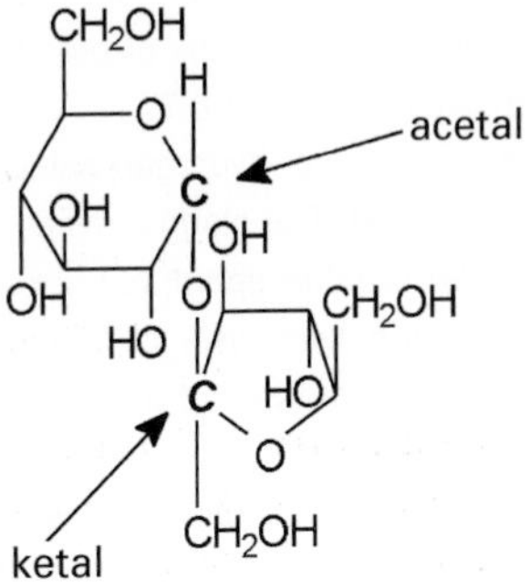

Acetals are formed when the anomeric carbon of aldoses become part of a glycosidic bond. The anomeric carbon of glucose (identified in bold) is part of a glycosidic bond, and is therefore an acetal. Similarly, ketals arise if the anomeric carbon of a ketose is used in glycosidic bond formation. This is the case for the fructoside part of sucrose.

(b) Aldoses and ketoses can reduce compounds when the carbonyl carbon is available. This is the case for open chain forms and cyclic forms that result in hemiacetal and hemiketal formation. Acetals and ketals are much more stable and are less prone to oxidation. Both anomeric carbons in sucrose are as ketals and acetals and so are not easily reduced. This makes sucrose an excellent storage sugar.

CHAPTER 8

Q8.1 (a) The nucleophilic, slightly negative site is the oxygen atom: $CH_3{-}\overset{\delta-}{O}{-}H$

(b) The free radical site is the second carbon with valency three: $CH_3{-}CH^{\cdot}{-}CH_2{-}CH_3$

(c) The electrophilic site, slightly positive, is the carbon atom bound to oxygen:

$$CH_3{-}\overset{\delta+}{\underset{\large |\atop H}{C}}{=}O$$

Q8.2 (a) Nucleophilic substitution is the replacement of an atom or group in a reactant molecule by a second group which attacks a nucleophilic, slightly positive, site in the reactant with a lone pair of electrons.

(b) The transition state in a reaction is the short-lived, high-energy stage between reactants and products when bonds are being formed or being broken. It is often designated by placing it in square brackets.

Q8.3 Electrophilic addition is important in the citric acid cycle where *cis*-aconitate is converted to isocitrate and fumarate is converted to malate by this reaction, the enzymes are aconitate and fumarase respectively.

One example from: *cis*-aconitate to isocitrate or fumarate to malate or alkyl halide to alkene. In the example, the reactants should be drawn in the appropriate orientation with partial charges; the curved arrows indicating the movement of pairs of electrons should be shown; the transition state should be indicated with partial bonds and placed in square brackets; the two stages of the mechanism should be clearly separated (see Figures 8.4 and 8.5).

Q8.4 (a) Two-stage elimination involves an initial step in which a base abstracts a proton from a carbon–hydrogen bond in the reactant to leave a carbanion. This is followed, in a separate step, by loss of an anion and formation of a carbon–carbon double bond. In concerted elimination, the two steps of proton abstraction and loss of an anion take place simultaneously (see Figure 8.7).

(b) Concerted elimination: loss of hydrogen halide from a haloalkane; loss of water from a β-hydroxycarboxylic acid. Two-stage elimination: loss of water from a β-hydroxyketone or β-hydroxythioester or other appropriate examples.

Q8.5 The chymotrypsin-mediated hydrolysis of a peptide adjacent to an aromatic amino acid; the papain-mediated hydrolysis of a peptide at a basic amino acid site; the base-catalysed hydrolysis of an ester; or another suitable example. The mechanism should show nucleophilic attack to form a charged intermediate followed by reformation of the carbonyl bond and bond clevage.

Q8.6 (a) FADH·

(b) The oxygen molecule has two unpaired electrons; it is a diradical. It can readily take up one or two electrons from FADH or one electron from FADH·. Thus FAD is a flexible agent for the transfer of electrons to oxygen.

Q8.7 (a) A reactive site is an atom or a bond within a molecule that provides a centre for chemical reactivity. The reactive site is usually the functional group with a polar bond.

(b) Lewis acid sites are slightly positive or electrophilic. They occur at an atom which is the positive end of a polar bond. Alcohol, amine, alkyl halide and carbonyl examples are given in Table 8.1. Lewis base sites are slightly negative or nucleophilic. They arise at the negative end of a polar covalent bond. Examples are listed in Table 8.1.

(i) The carbonyl group, —CO—, in butanone contains a slightly positive carbon atom (Lewis acid site) and a slightly negative oxygen (Lewis base).

(ii) 2-Aminopropane has the amine reactive site, $-NH_2$, with a slightly negative lone pair of electrons on nitrogen (Lewis base).

(iii) The reactive site in methyl ethanoate is the carbonyl group as detailed in answer (i).

Q8.8 Concerted nucleophilic substitution occurs in the reaction of an alkyl halide with a hydroxide ion or in the enzyme-mediated hydrolysis of glycogen, see Figures 8.2 and 8.3, respectively. Concerted elimination is found in organic chemistry with the elimination of hydrogen halide from an alkyl halide using an alkali or in the dehydration of β-hydroxycarboxylic acids using an enzyme catalyst (section 8.5, Figures 8.8 and 8.11).

Q8.9 Within the citric acid cycle, dehydration of citrate to form *cis*-aconitate is followed immediately by hydration of the product to give isocitrate. In a subsequent part of the cycle, dehydrogenation of succinate to fumarate is followed by hydration of fumarate to yield malate. Thus the link between the two processes is clear.

Q8.10 (a) Carbon–carbon bonds are formed in biosynthesis by: nucleophilic attack of a carbanion at an electrophilic centre such as a carbonyl group; electrophilic attack of a carbonium ion on an electron-rich alkene bond; by reaction of a free radical, on a carbon atom within a suitable molecule, with an alkene.

(b) An example can be taken from section 8.8.

CHAPTER 9

Q9.1 (a) 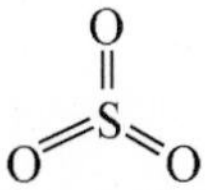 $O{=}S({=}O){=}O$ (b) Six. (c) 3s and 3p orbitals need unpairing for a valency of six.

Q9.2 (a) Valency of sulphur in coenzyme A is two. (b) Sulphur in sulphate has a valency of six.

Q9.3 (a) Enzymes are regulated by oxidation and reduction of disulphide bridges. Glutathione is an intermediate in reducing peroxides and is also reduced by glutathione reductase.

(b) Thioesters are more easily formed and are more reactive than organic esters, so that in the biological formation of esters, thioesters allow the overall process to take place in smaller steps.

Q9.4 This revolves around the numbering of the carbons. Identify the functional group that defines the numbering. This is the anomeric carbon (see answer to Question 7.8 for explanation) for (i) and the carboxyl in (ii). These are both carbon 1 and numbering is as follows:

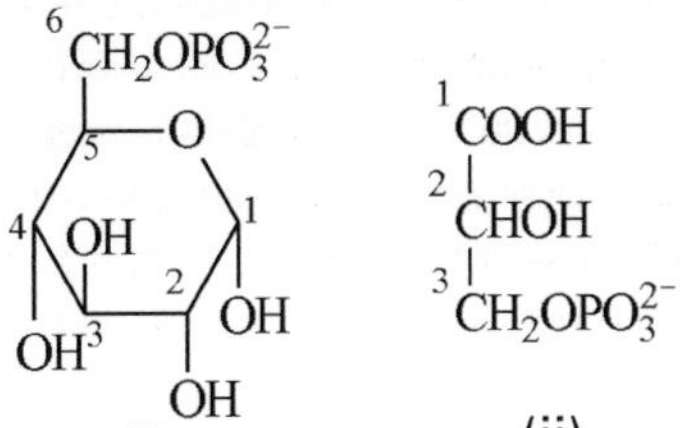

(i) Ring structure: ${}^6CH_2OPO_3^{2-}$ on C5; ring O; C1–OH; C2–OH; C3–OH; C4–OH

(ii) ${}^1COOH$ — ${}^2CHOH$ — ${}^3CH_2OPO_3^{2-}$

The phosphate ester is at carbon 6 for (i) and carbon 3 for (ii).

Q9.5 (a) Cyclic nucleotides, nucleic acids and phospholipids.

(b) The titratable oxygen has a p*K* of 1–2. It is thus dissociated and the conjugate base (anionic) predominates at pH 7.0.

Q9.6 (a) Reduced glutathione

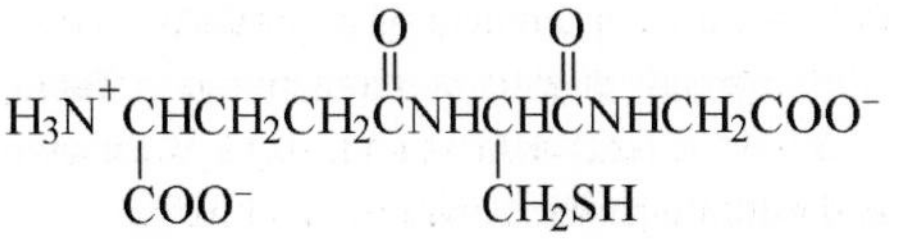

$$H_3\overset{+}{N}\underset{\displaystyle COO^-}{CH}CH_2CH_2\overset{O}{\overset{\|}{C}}NH\underset{\displaystyle CH_2SH}{CH}\overset{O}{\overset{\|}{C}}NHCH_2COO^-$$

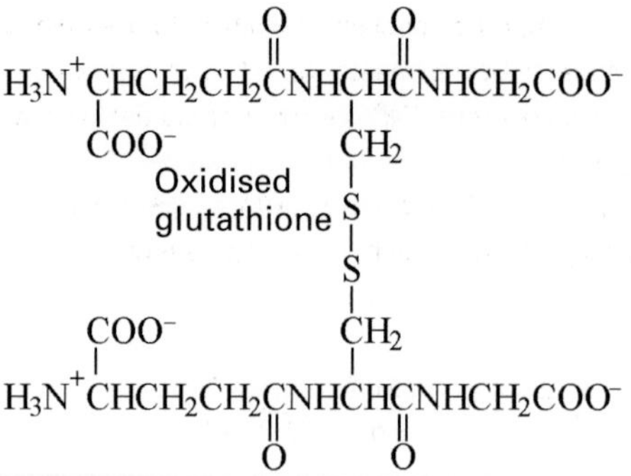

$$
\begin{array}{l}
\quad\quad\quad\quad\quad\quad\quad\quad\quad\ \mathrm{O}\quad\quad\quad\ \mathrm{O}\\
\quad\quad\quad\quad\quad\quad\quad\quad\quad\ \|\quad\quad\quad\ \ \|\\
\mathrm{H_3N^+CHCH_2CH_2CNHCHCNHCH_2COO^-}\\
\quad\quad\ \ |\quad\quad\quad\quad\quad\quad\quad\ |\\
\quad\quad\mathrm{COO^-}\quad\quad\quad\quad\ \mathrm{CH_2}\\
\quad\quad\quad\quad\quad\quad\quad\quad\quad\quad\ \ |\\
\quad\quad\quad\quad\quad\quad\quad\quad\quad\quad\ \mathrm{S}\\
\quad\quad\quad\quad\quad\quad\quad\quad\quad\quad\ \ |\\
\quad\quad\quad\quad\quad\quad\quad\quad\quad\quad\ \mathrm{S}\\
\quad\quad\quad\quad\quad\quad\quad\quad\quad\quad\ \ |\\
\quad\quad\mathrm{COO^-}\quad\quad\quad\quad\ \mathrm{CH_2}\\
\quad\quad\ \ |\quad\quad\quad\quad\quad\quad\quad\ |\\
\mathrm{H_3N^+CHCH_2CH_2CNHCHCNHCH_2COO^-}\\
\quad\quad\quad\quad\quad\quad\quad\quad\quad\ \|\quad\quad\quad\ \ \|\\
\quad\quad\quad\quad\quad\quad\quad\quad\quad\ \mathrm{O}\quad\quad\quad\ \mathrm{O}
\end{array}
$$

Oxidised glutathione

(b) The sulphur-containing functional group in reduced glutathione is the thiol, and the disulphide bridge in oxidised glutathione.

Q9.7 Phosphate is an excellent buffer at pH 7.2 because it has a p*K* at pH 7.0. Compounds are extremely good buffers near their p*K*.

Q9.8 (a)

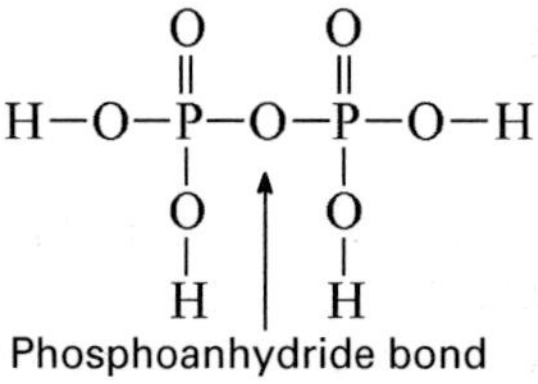

$$
\begin{array}{l}
\quad\quad\quad\ \ \mathrm{O}\quad\quad\ \ \mathrm{O}\\
\quad\quad\quad\ \ \|\quad\quad\quad \|\\
\mathrm{H{-}O{-}P{-}O{-}P{-}O{-}H}\\
\quad\quad\quad\ \ |\quad\quad\quad |\\
\quad\quad\quad\ \ \mathrm{O}\quad\quad\ \ \mathrm{O}\\
\quad\quad\quad\ \ |\quad\quad\quad |\\
\quad\quad\quad\ \ \mathrm{H}\quad\quad\ \ \mathrm{H}
\end{array}
$$

Phosphoanhydride bond

(b) Two from: electrostatic repulsion, phosphate formed by hydrolysis is stabilised by resonance, gain in entropy when hydrolysis takes place.

(c) Three from: temperature, concentration, pH, presence of metal ions, and enzymes.

Q9.9 (a) $ROH + R'OPO_3^{2-} \rightleftharpoons RO(R'O)PO_2^- + OH^-$

(b) Cellular messengers, storage and transmission of genetic codes, and polar group in phospholipids.

CHAPTER 10

Q10.1 (a) (i) Ethanal has been oxidised and NAD^+ reduced. (ii) Cytochrome c (Fe^{2+}) has been oxidised and Cu^{2+} has been reduced.

(b) (i) Ethanal is the reductant. (ii) Cytochrome c (Fe^{2+}) is the reductant.

Q10.2 The format for a general formula is $OX + ne^- + mH^+ \rightarrow RED$.

(i) $NAD^+ + 2e^- + 2H^+ \rightarrow NADH + H^+$. Malate oxidation is written opposite to the general formula and needs changing to: oxaloacetate + $2e^-$ + $2H^+$ → malate.

(ii) cytochrome c (Fe^{2+}) → cytochrome c (Fe^{3+}) + e^- and $Cu^{2+} + e^- \rightarrow Cu^+$. Cytochrome oxidation is written opposite to the general formula and needs changing to: cytochrome c (Fe^{3+}) + e^- → cytochrome c (Fe^{2+}).

Q10.3 (a) (i) Lactate/pyruvate and $NADH/NAD^+$ standard reduction potentials are –0.19 V and –0.32 V. Electrons will flow from negative to positive and will therefore go from NADH to pyruvate. (ii) $ascorbate_{(red)}/dehydroascorbate_{(ox)}$ and Fe^{2+}/Fe^{3+} standard reduction potentials are 0.08 V and 0.77 V. Electrons will flow from negative to positive and will therefore go from ascorbate to Fe^{3+} (b) (i) NADH will become oxidised and pyruvate reduced. (ii) Ascorbate will become oxidised and Fe^{3+} reduced.

Q10.4 (a) Split reactions into half-reactions: (i) $NAD^+ + 2e^- + 2H^+ \rightleftharpoons NADH + H^+$ and pyruvate + $2e^-$ + $2H^+$ → lactate; (ii) $Fe^{3+} \rightarrow Fe^{2+} + e^-$ and $Cu^{2+} + e^- \rightarrow Cu^+$.

(b) Look up values of $E°$ and the number of electrons transferred in Table 10.1 for the redox half-reactions. (i) –0.32 V for $NAD^+/NADH$ and –0.19 V for pyruvate/lactate; (ii) +0.34 V for Cu^{2+}/Cu^+ and +0.77 V for Fe^{3+}/Fe^{2+}.

(c) Calculate ΔE in the direction that the reaction is written. (i) Two electrons will flow from NADH to pyruvate and so NADH is the donor. ΔE is therefore $-0.19 - (-0.32) = +0.13$ V. (ii) One electron will flow from Fe^{2+} to Cu^{2+} and so Fe^{2+} is the donor. ΔE is therefore $+0.34 - (+0.77) = -0.43$ V.

(d) Substituting into $\Delta G° = -n\Delta EF$ gives (i) $\Delta G° = -2 \times +0.13 \times 96\,500$ J mol^{-1} = $-25\,090$ J mol^{-1}. (ii) $\Delta G° = -1 \times -0.43 \times 96\,500$ J mol^{-1} = $+41\,495$ J mol^{-1}.

Q10.5 (a) Use the Nernst equation

$$\Delta E = \Delta E° + \frac{0.06}{n} \log_{10} \frac{[\mathrm{Ox}]}{[\mathrm{Red}]}$$

to calculate the ΔE for pyruvate and lactate at given concentrations.

$$\Delta E = -0.19 + \frac{0.06}{2} \log_{10} \frac{(3 \times 10^{-3})}{(6 \times 10^{-4})}$$

$$= -0.19 + 0.02$$
$$= -0.17 \text{ V}$$

(b) Calculate ΔE. Two electrons will flow from NADH to pyruvate and so NADH is the donor. ΔE is therefore $-0.17 - (-0.32) = +0.15$ V.

(c) Substituting into $\Delta G° = -n\Delta EF$ gives $\Delta G° = -2 \times +0.15 \times 96\,500$ J mol^{-1} = $-28\,950$ J mol^{-1}.

Q10.6 Glutathione reductase catalyses electron transfer between glutathione and NADPH. The reaction can be summarised as:

$$\text{glutathione}_{(\mathrm{Ox})} + \text{NADPH} + \text{H}^+ \rightarrow 2\ \text{glutathione}_{(\mathrm{Red})} + \text{NADP}^+$$

(a) The standard reduction potentials are -0.23 V for glutathione and -0.32 V for NADPH. Electrons will flow from NADPH to glutathione.

(b) Calculate $\Delta E°$ for the direction written (in this case it will not matter if you calculate the reverse reaction). Glutathione is the acceptor and $\Delta E° = -0.23 - (-0.32) = +0.09$ V. Substituting into $\Delta G° = -n\Delta EF$ gives $\Delta G° = -2 \times +0.09 \times 96\,500$ J mol^{-1} = $-17\,370$ J mol^{-1}. If calculated the other way energy would be $+17\,370$ J mol^{-1}, which would be available for reaction if reversed.

(c) Use the Nernst equation

$$\Delta E = \Delta E° + \frac{0.06}{n} \log_{10} \frac{[\mathrm{Ox}]}{[\mathrm{Red}]}$$

to calculate the ΔE for glutathione at the given concentrations.

$$\Delta E = -0.23 + \frac{0.06}{2} \log_{10} \frac{(1)}{(500)}$$

$$= -0.23 - 0.08$$
$$= -0.31 \text{ V}$$

Then calculate ΔE for the direction written. Glutathione is the acceptor and $\Delta E° = -0.31 - (-0.32) = +0.01$ V. Substituting into $\Delta G° = -n\Delta EF$ gives $\Delta G° = -2 \times +0.01 \times 96\,500$ J mol^{-1} = -1930 J mol^{-1}.

Q10.7 In mitochondrial electron transport, electrons are transferred between ubiquinone and cytochrome c.

(a) Standard reduction potentials are +0.10 V for ubiquinone/ubiquinol and +0.22 V for cytochrome c. Electrons flow from ubiquinone to cytochrome c.

(b) Calculate the ΔE for the direction written. Cytochrome c is the acceptor and ubiquinol the donor and $\Delta E = +0.22 - (+0.10) = +0.12$ V. Substituting into $\Delta G = -n\Delta EF$ gives $\Delta G° = -2 \times 0.12 \times 96\,500 \text{ J mol}^{-1} = -23\,160 \text{ J mol}^{-1}$.

CHAPTER 11

Q11.1 (a) Suitable examples would be light bulbs or fluorescent tubes.

(b) During exercise, chemical energy is used to produce kinetic energy.

Q11.2 Heat taken up by the water = 13.5 × 500 × 4.18 J = 28 215 J

This was released by the combustion of 1.8 g glucose

Molar mass of glucose = 180 g mol^{-1}

$$1.8 \text{ g glucose} = \frac{1.8}{180} \text{ moles} = 0.01 \text{ moles}$$

$$\text{Heat produced by combustion of glucose} = \frac{28\,215}{0.01} \text{ J mol}^{-1} = 2\,821\,500 \text{ J mol}^{-1}$$

Converting to kJ mol^{-1} and rounding off to a sensible number of digits gives:

$$\text{Enthalpy of combustion of glucose} = 2822 \text{ kJ mol}^{-1}$$

Q11.3 The target equation is:

$$C_6H_{12}O_{6(s)} \rightarrow 2C_2H_5OH_{(l)} + 2CO_{2(g)}$$

The given equations are:

(1) $$C_6H_{12}O_{6(s)} + 6O_{2(g)} \rightarrow 6CO_{2(g)} + 6H_2O_{(l)}$$

$$\Delta H_1 = -2821 \text{ kJ mol}^{-1}$$

(2) $$C_2H_5OH_{(l)} + 3O_{2(g)} \rightarrow 2CO_{2(g)} + 3H_2O_{(l)}$$

$$\Delta H_2 = -1368 \text{ kJ mol}^{-1}$$

Equation (1) contains glucose, $C_6H_{12}O_{6(s)}$, on the correct side and in the correct numbers to match the target equation. This equation therefore requires no manipulation.

Equation (2) contains ethanol, $C_2H_5OH_{(l)}$, on the wrong side and in the wrong numbers to match the target equation. To get the ethanol on the correct side, we reverse equation (2):

(3) $$2CO_{2(g)} + 3H_2O_{(l)} \rightarrow C_2H_5OH_{(l)} + 3O_{2(g)}$$

$$\Delta H_3 = +1368 \text{ kJ mol}^{-1}$$

To obtain the correct number of ethanol molecules, we multiply equation (3) by 2:

(4) $$4CO_{2(g)} + 6H_2O_{(l)} \rightarrow 2C_2H_5OH_{(l)} + 6O_{2(g)}$$

$$\Delta H_4 = 2 \times (+1368) = +2736 \text{ kJ mol}^{-1}$$

In equations (1) and (4) glucose and ethanol are present in the correct positions and in the correct numbers to match the target equation. We now add these two equations:

$$C_6H_{12}O_{6(s)} + 6O_{2(g)} + 4CO_{2(g)} + 6H_2O_{(l)} \rightarrow 6CO_{2(g)} + 6H_2O_{(l)} + 2C_2H_5OH_{(l)} + 6O_{2(g)}$$

Cancelling out the species that occur on both sides of the equation leaves us with the target equation:

(5) $$C_6H_{12}O_{6(s)} \rightarrow 2C_2H_5OH_{(l)} + 2CO_{2(g)}$$

$$\Delta H_5 = (-2821) + (+2736)$$
$$= -85 \text{ kJ mol}^{-1}$$

The energy available from the anaerobic respiration of glucose = –85 kJ mol^{-1}.

Q11.4 Using the equation:

$$\Delta H_{reaction} = \Sigma\ \Delta H_f(\text{products}) - \Sigma\ \Delta H_f(\text{reactants})$$

and substituting in the given values (remembering that the enthalpy of formation of any element in its standard state is zero) gives:

$$\Sigma\ \Delta H_f(\text{products}) = -484.5 + (-285.5)$$
$$= -770 \text{ kJ mol}^{-1}$$

$$\Sigma\ \Delta H_f(\text{reactants}) = -277.7 + 0 = -277.7 \text{ kJ mol}^{-1}$$

$$\Delta H_{reaction} = -770 - (-277.7)$$
$$= -492.3 \text{ kJ mol}^{-1}$$

Acetobacter can therefore extract –492.3 kJ mol^{-1} by oxidising ethanol in this way.

Q11.5 Using the equation:

$$\Delta G = \Delta H - T\Delta S$$

and substituting in the given values gives:

$$-3700 = 14\,900 - 298\Delta S$$

$$298\Delta S = 14\,900 - (-3700)$$
$$= 18\,600 \text{ J mol}^{-1}$$

$$\Delta S = \frac{18\,600}{298}$$

$$= 62.4 \text{ J K}^{-1} \text{ mol}^{-1}$$

The entropy change involved in the conversion of fumarate to malate at 25°C = 62.4 J K^{-1} mol^{-1}.

Q11.6 (a) The first law of thermodynamics states that energy cannot be created or destroyed, but only converted from one form to another.

(b) The SI unit of energy is the joule (J).

Q11.7 (a) An exothermic reaction is one which releases heat to the surroundings.
(b) An endothermic reaction is one which absorbs heat from the surroundings.
If the reaction vessel gets hot during the course of a reaction, the reaction is exothermic.

Q11.8 The target equation is:

$$C_6H_{12}O_6 + O_2 \rightarrow 2C_3H_4O_3 + 2H_2O$$

The given equations are:

(1) $$C_6H_{12}O_6 + 6O_2 \rightarrow 6CO_2 + 6H_2O$$

$$\Delta H_{comb} = -2822 \text{ kJ mol}^{-1}$$

(2) $$C_3H_4O_3 + \tfrac{5}{2}O_2 \rightarrow 3CO_2 + 2H_2O$$

$$\Delta H_{comb} = -1168 \text{ kJ mol}^{-1}$$

Equation (1) has glucose, $C_6H_{12}O_6$, on the correct side and in the correct numbers to match the target equation. This equation will not require any manipulation.
Equation (2) has the pyruvic acid, $C_3H_4O_3$, on the wrong side and in the wrong numbers to match the target equation.
To get the pyruvic acid on the correct side, we reverse equation (2):

(3) $$3CO_2 + 2H_2O \rightarrow C_3H_4O_3 + \tfrac{5}{2}O_2$$

$$\Delta H = +1168 \text{ kJ mol}^{-1}$$

To obtain the correct number of pyruvic acid molecules, we multiply equation (3) by 2:

(4) $$6CO_2 + 4H_2O \rightarrow 2C_3H_4O_3 + 5O_2$$

$$\Delta H = 2\ (+1168) = +2336 \text{ kJ}$$

Equations (1) and (4) contain glucose and pyruvic acid on the correct sides and in the correct numbers to match the target equation. We now add these two equations:

$$C_6H_{12}O_6 + 6O_2 + 6CO_2 + 4H_2O \rightarrow 6CO_2 + 6H_2O + 2C_3H_4O_3 + 5O_2$$

Cancelling out species which occur on both sides of this equation leaves us with the target equation:

$$C_6H_{12}O_6 + O_2 \rightarrow 2C_3H_4O_3 + 2H_2O$$

$$\Delta H = -2822 + (+2336)$$
$$= -486 \text{ kJ mol}^{-1}$$

Enthalpy released in this part of the TCA cycle = -486 kJ mol^{-1}.

Q11.9 The reaction involved is:

$$\text{fumaric acid} \rightarrow \text{maleic acid}$$

Using the equation:

$$\Delta H_{reaction} = \Sigma\ \Delta H_f(\text{products}) - \Sigma\ \Delta H_f(\text{reactants})$$

and substituting in the given values gives:

$$\begin{aligned}\Delta H_{reaction} &= -785 - (-810)\\ &= +25\ \text{kJ mol}^{-1}\end{aligned}$$

The enthalpy of isomerisation of fumaric to maleic acid = +25 kJ mol^{-1}.

CHAPTER 12

Q12.1 (a) Using the equation:

$$\Delta G = -RT \ln K$$

and substituting in the given values gives:

$$\begin{aligned}13\,000 &= -8.314 \times 310 \times \ln K\\ &= -2577.34 \times \ln K\end{aligned}$$

$$\ln K = \frac{13\,000}{-2577.34}$$

$$= -5.04$$

$$K = 0.00645$$

The equilibrium constant for this reaction = 0.00645.

(b) The value of the equilibrium constant calculated in (a) shows that the reactants, leucine and glycine, will predominate in the equilibrium mixture.

Q12.2 Using the equation:

$$\ln\left(\frac{\text{Rate}_1}{\text{Rate}_2}\right) = -\frac{E_a}{R}\left(\frac{1}{T_1} - \frac{1}{T_2}\right)$$

and substituting in the given values gives:

$$\ln\left(\frac{1.24 \times 10^{-3}}{3.16 \times 10^{-3}}\right) = -\frac{E_a}{8.314}\left(\frac{1}{295} - \frac{1}{303}\right)$$

$$-0.935 = -\frac{E_a}{8.314}\left(8.95 \times 10^{-5}\right)$$

$$E_a = \frac{0.935 \times 8.314}{8.95 \times 10^{-5}}$$

$$= 86\,855.8\ \text{J mol}^{-1}$$

Converting to kJ and rounding off to a sensible number of digits gives:

$$\text{Activation energy} = 86.9\ \text{kJ mol}^{-1}$$

Q12.3 A graph of $\frac{1}{[\text{Penicillin}]}$ versus $\frac{1}{\text{Initial rate}}$ must be plotted. First, the correct values must be calculated:

$\frac{1}{[\text{Penicillin}]}/10^4\ \text{mol}^{-1}\ \text{dm}^3$	$\frac{1}{\text{Initial rate}}/10^7\ \text{mol}^{-1}\ \text{dm}^3\ \text{min}$
100	9.09
33.3	4.00
20	2.94
10	2.22
3.3	1.72
2	1.64

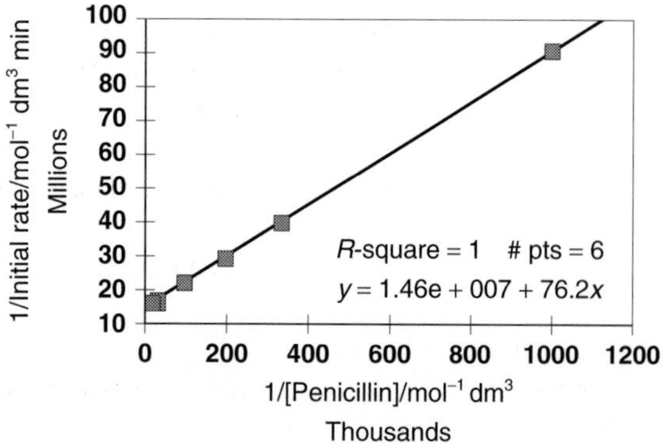

The graph tells us that the intercept = 1.46×10^7 and the gradient = 76.2.

$$\frac{1}{V_{\text{max}}} = 1.46 \times 10^7$$

$$V_{\text{max}} = \frac{1}{1.46 \times 10^7}$$

$$= 6.85 \times 10^{-8}\ \text{mol dm}^{-3}\ \text{min}^{-1}$$

$$\frac{K_{\text{m}}}{V_{\text{max}}} = 76.2$$

$$K_{\text{m}} = 76.2 \times V_{\text{max}}$$
$$= 76.2 \times 6.85 \times 10^{-8}$$
$$= 5.22 \times 10^{-6}\ \text{mol dm}^{-3}$$

Therefore, for penicillinase under these conditions, $V_{\text{max}} = 6.85 \times 10^{-8}$ mol dm^{-3} min^{-1} and $K_{\text{m}} = 5.22 \times 10^{-6}$ mol dm^{-3}.

Q12.4 Using the equation:

$$\Delta G = -RT \ln K_{\text{eq}}$$

and substituting in the given values of ΔG, R and T gives:

$$-30\,900 = -8.314 \times 310 \times \ln K_{\text{eq}}$$
$$= -2577.34 \times \ln K_{\text{eq}}$$

$$\ln K_{eq} = \frac{-30\,900}{-2577.34}$$

$$= 11.99$$

$$K_{eq} = 1.61 \times 10^5$$

The equilibrium constant between ATP and ADP under these conditions = 1.61×10^5.

Q12.5 Using the equation:

$$\Delta G = -RT \ln K_{eq}$$

and substituting in the given values of R, T and K_{eq} gives:

$$\Delta G = -8.314 \times 310 \times \ln 19$$
$$= -8.314 \times 310 \times 2.944$$
$$= -7588.8 \text{ J mol}^{-1}$$

Converting to kJ mol^{-1} and rounding off to a sensible number of digits gives:

$$\Delta G = -7.59 \text{ kJ mol}^{-1}$$

Q12.6 Raising the temperature has two effects:

(i) it raises the average speed of the molecules, increasing the violence and frequency of their collisions;

(ii) it increases the number of molecules travelling at speeds much higher than average.

The more violent the collisions, the more energy is available to activate the molecules. The more frequent the collisions, the more chances the molecules have of reacting.

Q12.7 Using the equation:

$$\ln\left(\frac{\text{Rate}_1}{\text{Rate}_2}\right) = -\frac{E_a}{R}\left(\frac{1}{T_1} - \frac{1}{T_2}\right)$$

and substituting in the given values gives:

$$\ln\left(\frac{6.1 \times 10^{-5}}{3.2 \times 10^{-4}}\right) = -\frac{E_a}{8.314}\left(\frac{1}{298} - \frac{1}{310}\right)$$

$$-1.657 = -\frac{E_a}{8.314}\left(1.299 \times 10^{-4}\right)$$

$$E_a = -\frac{1.657 \times 8.314}{1.299 \times 10^{-4}}$$

$$= 106\,053.1 \text{ J mol}^{-1}$$

Converting to kJ mol^{-1} and rounding off to a sensible number of digits gives:

$$\text{Activation energy} = 106.1 \text{ kJ mol}^{-1}$$

Q12.8 See Figure 12.4.

Q12.9 A graph of $\frac{1}{[\text{Isocitrate}]}$ versus $\frac{1}{\text{Initial rate}}$ must be plotted. First, the correct values must be calculated:

$\frac{1}{[\text{Isocitrate}]}/10^4\ \text{mol}^{-1}\ \text{dm}^3$	$\frac{1}{\text{Initial rate}}/10^8\ \text{mol}^{-1}\ \text{dm}^3\ \text{min}$
10	3.50
5.0	2.38
3.3	2.00
2.5	1.81
2.0	1.70

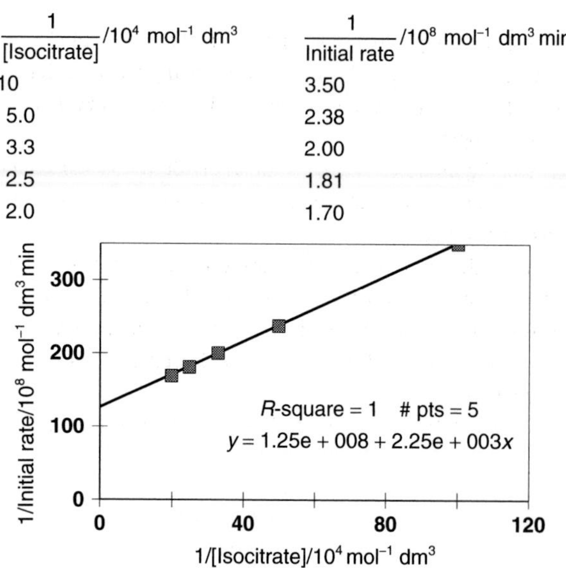

The graph tells us that the intercept = 1.25×10^8 and the gradient = 2.25×10^3.

$$\frac{1}{V_{max}} = 1.25 \times 10^8$$

$$V_{max} = \frac{1}{1.25 \times 10^8}$$

$$= 8.00 \times 10^{-9}\ \text{mol dm}^{-3}\ \text{min}^{-1}$$

$$\frac{K_m}{V_{max}} = 2.25 \times 10^3$$

$$K_m = 2.25 \times 10^3 \times V_{max}$$
$$= 1.8 \times 10^{-5}\ \text{mol dm}^{-3}$$

CHAPTER 13

Q13.1 A wavelength is the shortest distance between two similar points in successive waves. In this figure these are lines **a** and **c**.

Q13.2 254 nm is 2.54×10^{-7} m. The equation required is $v\lambda = c$. This needs transfoming to isolate v. Dividing both sides by λ and cancelling out terms gives $v = c/\lambda$. $c = 3 \times 10^8\ \text{m s}^{-1}$. Substitute in values to give

$$v = \frac{3 \times 10^8}{2.54 \times 10^{-7}}$$

$$= 1.18 \times 10^{15}\ \text{s}^{-1}$$

Q13.3 (a) 198 nm is 1.98×10^{-7} m. Substitute into $E = \frac{hc}{\lambda}$, $c = 3 \times 10^8\ \text{m s}^{-1}$, $h = 6.62 \times 10^{-34}$ J S

$$E = \frac{(6.64 \times 10^{-34}) \times (3 \times 10^{8})}{1.98 \times 10^{-7}}$$

$$= 1.01 \times 10^{-18}\ \text{J}$$

(b) The energy of a photon of 198 nm is greater than that of a photon of 550 nm.

Q13.4 Count the number of double bonds that are conjugated in each compound. Compound (b) has the highest number of conjugated double bonds and therefore absorbs light at the highest wavelength.

Q13.5 (a) Examine Figure 13.6. The wavelength of 450 nm would be the best as chlorophyll b has the highest absorbance at this wavelength. Large absorbances are easier to measure accurately.

(b) You must choose a wavelength where the absorbance of chlorophyll b is high and, ideally, there is no absorbance by chlorophyll a. This is not necessarily the same as the best absorbance for measuring the pure compound. In this instance 450 nm is also a good wavelength to choose as chlorophyll b has a high absorbance relative to chlorophyll a.

Q13.6 Use the Beer–Lambert law $A = \varepsilon cl$. Then substitute in terms.

$$A = \varepsilon cl$$
$$= 5400 \times 10^{-4} \times 1$$
$$= 0.54$$

Q13.7 Method as in answer to Question 13.4(b). Answer is compound (c).

Q13.8 Transform the Beer–Lambert law to isolate ε. Divide both sides by *cl* and cancel out terms. Now substitute in given values:

$$\frac{A}{cl} = \frac{\varepsilon cl}{cl}$$

$$\varepsilon = \frac{A}{cl} = \frac{0.3}{0.3 \times 1} = 1\ \text{cm}^{-1}$$

Q13.9 (a) The reflecting prism splits the light source into component wavelengths by diffraction. This then allows you to select light of a specific wavelength.

(b) The photocell converts light intensity to an electrical signal that can be transformed.

(c) The variable slit controls the amount of light entering the sample in the cuvette.

APPENDIX: DERIVATIONS OF EQUATIONS

DERIVATION 5.1

THE RELATIONSHIP BETWEEN pK_a, pK_b AND pK_w

In Chapter 5 we saw that:

$$K_a K_b = K_w$$

Taking logarithms to base 10 gives:

$$\log_{10} K_a + \log_{10} K_b = \log_{10} K_w$$

Multiplying through by −1 gives:

$$-\log_{10} K_a - \log_{10} K_b = -\log_{10} K_w$$

Since

$$pK_a = -\log_{10} K_a$$

$$pK_b = -\log_{10} K_b$$

and

$$pK_w = -\log_{10} K_w$$

$$pK_a + pK_b = pK_w = 14$$

DERIVATION 5.2

pH OF SOLUTIONS OF STRONG ACIDS

A strong acid is one which is fully dissociated in solution:

$$HX + H_2O \rightarrow H_3O^+ + X^-$$

Every molecule of acid produces a hydrogen ion when it dissociates. The concentration of hydrogen ions is therefore the same as the concentration of acid added, *C*.

By definition,

$$pH = -\log_{10}[H_3O^+]$$

So for a strong acid

$$pH = -\log_{10} C$$

DERIVATION 5.3

pH OF SOLUTIONS OF STRONG BASES

A strong base is one which is fully dissociated in solution:

$$MOH \rightarrow M^+ + OH^-$$

Each molecule of base produces a hydroxide ion when it dissociates. The concentration of hydroxide ion is therefore the same as the concentration of the base added, *C*.
By definition,

$$pOH = -\log_{10}[OH^-]$$

$$pOH = -\log_{10} C$$

Since

$$pH + pOH = pK_w$$

$$pOH = pK_w - pH$$

Therefore,

$$pK_w - pH = -\log_{10} C$$

$$pH = pK_w + \log_{10} C$$

DERIVATION 5.4

pH OF SOLUTIONS OF WEAK ACIDS

Weak acids only partially dissociate in water:

$$HA + H_2O \rightleftharpoons H_3O^+ + A^-$$

The equilibrium is represented by the acid dissociation constant, K_a.
For weak acids, we have:

$$K_a = \frac{[A^-][H_3O^+]}{[HA]}$$

For each hydrogen ion formed by the dissociation of the acid, a corresponding anion must be formed, so we have:

$$[H_3O^+] = [A^-]$$

so that:

$$K_a = \frac{[H_3O^+]^2}{[HA]}$$

Therefore,

$$[H_3O^+] = (K_a[HA])^{\frac{1}{2}}$$

and, since the acid is only slightly dissociated,

$$[HA] \cong C$$

$$\begin{aligned} pH &= -\log_{10} (K_aC)^{\frac{1}{2}} \\ &= \tfrac{1}{2}pK_a - \tfrac{1}{2}\log_{10} C \end{aligned}$$

DERIVATION 5.5

pH OF SOLUTIONS OF WEAK BASES

For weak bases, we have:

$$K_b = \frac{[OH^-][BH^+]}{[B]}$$

By similar reasoning to that in Derivation 5.4, we obtain:

$$[OH^-] = [BH^+]$$

$$[B] = C$$

$$K_b = \frac{[OH^-]^2}{C}$$

$$[OH^-] = (K_bC)^{\frac{1}{2}}$$

$$pOH = \tfrac{1}{2}pK_b - \tfrac{1}{2}\log_{10} C$$

$$pOH = pK_w - pH$$

$$pK_w - pH = \tfrac{1}{2}pK_b - \tfrac{1}{2}\log_{10} C$$

$$pH = pK_w - \tfrac{1}{2}pK_b + \tfrac{1}{2}\log_{10} C \qquad \text{(A5.1)}$$

In many cases, pK_a values are listed for bases as well as for acids. Since:

$$pK_a + pK_b = pK_w$$

$$pK_b = pK_w - pK_a$$

Substituting this in Equation A5.1 gives:

$$\begin{aligned} pH &= pK_w - \tfrac{1}{2}(pK_w - pK_a) + \tfrac{1}{2}\log_{10} C \\ &= pK_w - \tfrac{1}{2}pK_w + \tfrac{1}{2}pK_a + \tfrac{1}{2}\log_{10} C \end{aligned}$$

$$pH = \tfrac{1}{2}pK_w + \tfrac{1}{2}pK_a + \tfrac{1}{2}\log_{10} C$$

DERIVATION 5.6

pH OF SOLUTIONS OF SALTS OF STRONG BASES WITH WEAK ACIDS

The anion derived from the weak acid undergoes hydrolysis in water:

$$A^- + H_2O \rightleftharpoons AH + OH^-$$

K_b for this reaction is given by:

$$K_b = \frac{[AH][OH^-]}{[A^-]}$$

Since, from the reaction equation,

$$[OH^-] = [AH]$$

we can write:

$$K_b = \frac{[OH^-]^2}{[A^-]}$$

If the degree of hydrolysis is small (i.e. if K_b is small), then $[A^-]$ is almost the same as the concentration of the salt originally added, C:

$$K_b = \frac{[OH^-]^2}{C}$$

$$[OH^-] = (K_b C)^{\frac{1}{2}}$$

Since

$$[H_3O^+] = \frac{K_w}{[OH^-]}$$

We have:

$$[H_3O^+] = \frac{K_w}{(K_b C)^{\frac{1}{2}}}$$

$$-\log_{10}[H_3O^+] = -\log_{10}\frac{K_w}{(K_b C)^{\frac{1}{2}}}$$

$$pH = -\log_{10} K_w - (-\log_{10}(K_b C)^{\frac{1}{2}})$$

$$pH = pK_w - \tfrac{1}{2}pK_b + \tfrac{1}{2}\log_{10} C$$

Often, only pK_a values for the parent acid will be available. In this case,

$$pH = \tfrac{1}{2}pK_w + \tfrac{1}{2}pK_a + \tfrac{1}{2}\log_{10} C$$

DERIVATION 5.7

pH OF SOLUTIONS OF SALTS OF WEAK BASES WITH STRONG ACIDS

The cation derived from the weak base undergoes hydrolysis:

$$B^+ + H_2O \rightleftharpoons BOH + H^+$$

The solution is therefore acidic, with

$$K_a = \frac{[BOH][H^+]}{[B^+]}$$

Since, from the reaction equation,

$$[H^+] = [BOH]$$

we can write:

$$K_a = \frac{[H^+]^2}{[B^+]}$$

If the degree of hydrolysis is small (i.e. if K_a is small), then $[B^+]$ is almost the same as the concentration of the salt originally added, C:

$$K_a = \frac{[H^+]^2}{C}$$

Therefore,

$$[H_3O^+] = (K_aC)^{\frac{1}{2}}$$

$$-\log_{10}[H_3O^+] = -\log_{10}(K_aC)^{\frac{1}{2}}$$

$$pH = \tfrac{1}{2}pK_a - \tfrac{1}{2}\log_{10} C$$

If only pK_b data are available, the value for the parent base of the salt must be used, and the equation becomes:

$$pH = \tfrac{1}{2}pK_w - \tfrac{1}{2}pK_b - \tfrac{1}{2}\log_{10} C$$

DERIVATION 5.8

pH OF BUFFER SOLUTIONS

In a buffer prepared from a weak acid and one of its salts, the concentration of the anion derived from the weak acid will be equal to the concentration of the salt, since the dissociation of the acid is very slight:

$$[A\ ] \quad [salt]_o$$

where $[salt]_o$ is the concentration of salt added originally.

Similarly, the concentration of undissociated acid will be equal to the concentration of acid added:

$$[AH] \quad [acid]_o$$

where $[\text{acid}]_o$ is the concentration of acid added originally.
Now,

$$K_a = \frac{[A^-][H^+]}{[AH]}$$

$$= \frac{[\text{salt}]_o[H^+]}{[\text{acid}]_o}$$

$$[H^+] = \frac{K_a[\text{acid}]_o}{[\text{salt}]_o}$$

or

$$\text{pH} = \text{p}K_a + \log_{10}\frac{[\text{salt}]_o}{[\text{acid}]_o}$$

For buffers made with weak base and a strong-acid salt of the base, a similar line of reasoning leads to the formula:

$$\text{pH} = \text{p}K_a - \log_{10}\frac{[\text{salt}]_o}{[\text{base}]_o}$$

These equations can be combined into one:

$$\text{pH} = \text{p}K_a + \log_{10}\frac{[\text{unprotonated species}]}{[\text{protonated species}]}$$

DERIVATION 10.1

DERIVATION OF EQUATION $\Delta G° = -n\Delta E°F$ (A10.1)

The amount of work (w) that 1 mole of reactants and products can produce is equivalent to the free energy that can be obtained from the system or:

$$\Delta G° = w$$

For a reaction that transfers n moles of electrons at an electrochemical potential difference ΔE the amount of work that can be obtained is:

$$w = -n\Delta E°F$$

where F is the Faraday constant, equal to 96 485 C mol^{-1}. F is calculated from the electrical charge of 1 mole of electrons.
As $\Delta G° = w$ it also equals $= -n\Delta E°F$.

DERIVATION 10.2

THE NERNST EQUATION

The derivation begins with the following equation:

$$\Delta G = \Delta G° + 2.3RT \log_{10}\frac{[P]^p}{[S]^s} \quad \text{(A10.2)}$$

for the reaction of $sS \rightleftharpoons pP$. Equation A10.1 can be converted to the following form: $-\frac{\Delta G}{nF} = \Delta E$ and so dividing Equation A10.2 by $-nF$ converts the ΔG to ΔE as given:

$$\frac{\Delta G}{-nF} = \frac{\Delta G^\circ}{-nF} + \frac{2.3RT}{-nF} \log_{10} \frac{[P]^p}{[S]^s}$$

which converts to:

$$\Delta E = \Delta E^\circ - \frac{2.3RT}{nF} \log_{10} \frac{[P]^p}{[S]^s} \quad \text{(A10.3)}$$

The reaction $OX + ne^- \rightarrow RED$ can now be used in Equation A10.3 to yield:

$$\Delta E = \Delta E^\circ - \frac{2.3RT}{nF} \log_{10} \frac{[RED]}{[OX]}$$

which converts to the Nernst equation:

$$\Delta E = \Delta E^\circ + \frac{2.3RT}{nF} \log_{10} \frac{[OX]}{[RED]}$$

DERIVATION 12.1

THE MICHAELIS-MENTEN EQUATION

The proposed mechanism for enzyme action is:

$$E + S \underset{k_{-1}}{\overset{k_1}{\rightleftharpoons}} ES \overset{k_2}{\rightarrow} E + P$$

where E = enzyme, S = substrate, ES = enzyme–substrate complex and P = product. Also shown in the equation are the rate constants for the three possible reactions. Considering the formation of product, we can see that its rate of formation is given by:

$$\text{Rate of formation of product} = k_2[ES]$$

Unfortunately, we do not know the concentration of the enzyme–substrate complex, so this equation is of no use. We need to replace [ES] by some other expression. The enzyme–substrate complex is produced by the forward reaction, rate constant k_1, and destroyed either by the backward reaction, rate constant k_{-1}, or by proceeding to form product in the reaction with rate constant k_2. The net rate of production of ES will therefore be given by:

$$\text{Net rate of formation of ES} = k_1[E][S] - k_{-1}[ES] - k_2[ES]$$

When the reaction first starts, there is no ES present. Its concentration increases rapidly at the start of the reaction. As its concentration increases, so does the rate at which it breaks up, either to give enzyme and substrate back again or to give enzyme and product. At some point, the rates of formation and destruction of ES will be equal. The concentration of the enzyme–substrate complex will then not change until the very end of the reaction. It can therefore be assumed that the concentration of ES is constant during most of the time the reaction is proceeding. This is called the **steady state assumption**.

If we make this assumption, then the net rate of formation of ES is zero. Thus,

$$0 = k_1[\mathrm{E}][\mathrm{S}] - k_{-1}[\mathrm{ES}] - k_2[\mathrm{ES}]$$

Here, [E] is the concentration of free enzyme. Again, this is something we do not know, so the equation is not yet usable. We do know the total amount of enzyme we added, $[\mathrm{E}]_o$, and that:

$$[\mathrm{E}]_o = [\mathrm{E}] + [\mathrm{ES}]$$

$$[\mathrm{E}] = [\mathrm{E}]_o - [\mathrm{ES}]$$

$$\begin{aligned} 0 &= k_1([\mathrm{E}]_o - [\mathrm{ES}])[\mathrm{S}] - k_{-1}[\mathrm{ES}] - k_2[\mathrm{ES}] \\ &= k_1[\mathrm{E}]_o[\mathrm{S}] - k_1[\mathrm{ES}][\mathrm{S}] - k_{-1}[\mathrm{ES}] - k_2[\mathrm{ES}] \end{aligned}$$

Rearranging gives:

$$\begin{aligned} k_1[\mathrm{E}]_o[\mathrm{S}] &= k_1[\mathrm{ES}][\mathrm{S}] + k_{-1}[\mathrm{ES}] + k_2[\mathrm{ES}] \\ &= (k_1[\mathrm{S}] + k_{-1} + k_2)[\mathrm{ES}] \end{aligned}$$

Rearranging gives:

$$[\mathrm{ES}] = \frac{k_1[\mathrm{E}]_o[\mathrm{S}]}{(k_{-1} + k_2 + k_1[\mathrm{S}])}$$

We can now substitute this expression for [ES] in the equation for the rate of formation of product:

$$\text{Rate of formation of product} = \frac{k_1k_2[\mathrm{E}]_o[\mathrm{S}]}{(k_{-1} + k_2 + k_1[\mathrm{S}])}$$

Dividing top and bottom by k_1 gives:

$$\begin{aligned} \text{Rate of formation of product} &= \frac{k_2[\mathrm{E}]_o[\mathrm{S}]}{\left(\dfrac{k_{-1} + k_2}{k_1} + [\mathrm{S}]\right)} \\ &= \frac{k_2[\mathrm{E}]_o[\mathrm{S}]}{(K_m + [\mathrm{S}])} \end{aligned}$$

where $\dfrac{k_{-1} + k_2}{k_1} = K_m$

The maximum rate will occur when the enzyme is fully saturated with the substrate. $k_2[\mathrm{E}]_o$ will then be at its maximum value, V_{max}, and the equation becomes:

$$\text{Rate of formation of product} = \frac{V_{max}[\mathrm{S}]}{(K_m + [\mathrm{S}])}$$

DERIVATION 13.1

THE BEER–LAMBERT LAW

Consider the following system of a solution of absorbing species at concentration c in a transparent cell of path length l.

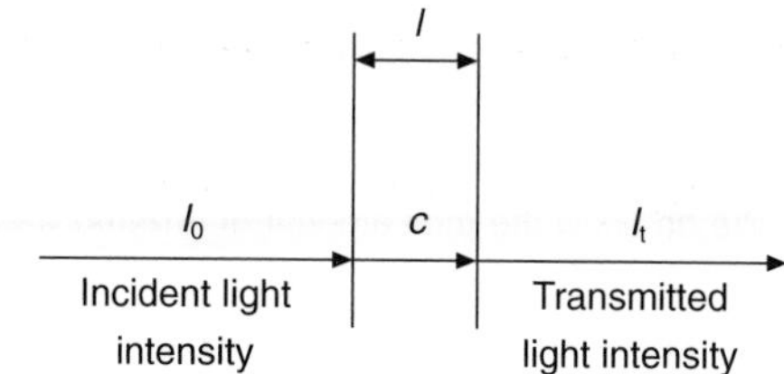

The light entering the cell has intensity I_0 whilst the light leaving the cell is at intensity I_t, some having been absorbed by the molecules in the cell.

Let us consider the absorption of photons of light by the chromophore (coloured compound) in solution. The number of photons absorbed will be proportional to the number of chromophore molecules that the light path encounters times the probability of that molecule absorbing a photon. The number of photons absorbed is proportional to the energy (and therefore the intensity of the transmitted light), whilst for any set compound in a set environment at a set wavelength the number of photons will be directly proportional to the concentration in solution. We can express this as an equation as follows:

$$dI = -kcI\,dl$$

This can be rearranged to give:

$$\frac{dI}{I} = -kc\,dl$$

where k is a constant and dI is the change in light intensity for an infinitesimally small thickness of sample dl. The negative sign indicates that the transmitted light intensity falls as the thickness increases. The absorbed light for all the thickness of sample can be calculated by integrating both sides:

$$\int_{I_0}^{I} \frac{dI}{I} = -kc \int_0^l dl$$

$$\frac{I_t}{I_0} = e^{-\varepsilon' cl}$$

where ε' is a new constant which takes into consideration molecule and medium.

Taking natural logarithms of both sides gives:

$$\ln \frac{I_t}{I_0} = -\varepsilon' cl$$

and rearranging:

$$\ln \frac{I_0}{I_t} = \varepsilon' cl$$

or:

$$\log_{10}\frac{I_0}{I_t} = \varepsilon c l$$

where the constant ε′ changes to ε, the molar absorptivity, to take account of the change in the base of the logarithm.

Substitution of A for the value $\log_{10}\frac{I_0}{I_t}$ gives the Beer–Lambert law.

INDEX

Note: page numbers in italics denote figures and tables; page numbers in bold denote Advanced Topic Boxes.